Democratic Experiments

Inside Technology
Edited by Wiebe E. Bijker, W. Bernard Carlson, and Trevor Pinch

A list of titles in the series appears at the back of the book.

Democratic Experiments

Problematizing Nanotechnology and Democracy in Europe and the United States

Brice Laurent

The MIT Press
Cambridge, Massachusetts
London, England

This book was set in Stone Sans and Stone Serif by Toppan Best-set Premedia Limited. Printed and bound in the United States of America.

Library of Congress Cataloging-in-Publication Data

Names: Laurent, Brice, author.
Title: Democratic experiments : problematizing nanotechnology and democracy in Europe and the United States / Brice Laurent.
Description: Cambridge, MA : MIT Press, [2017] | Series: Inside technology | Includes bibliographical references and index.
Identifiers: LCCN 2016028724 | ISBN 9780262035767 (hardcover : alk. paper)
Subjects: LCSH: Nanotechnology--Political aspects--European Union countries. | Nanotechnology--Political aspects--United States. | Nanotechnology--Government policy--European Union countries. | Nanotechnology--Government policy--United States. | Science and state--European Union countries. | Science and state--United States. | Democracy--European Union countries. | Democracy--United States.
Classification: LCC T174.7 .L379 2017 | DDC 338.94/06—dc23 LC record available at https://lccn.loc.gov/2016028724

10 9 8 7 6 5 4 3 2 1

Contents

Acknowledgments

This book has been developed over the past few years, while I was a PhD candidate and then a permanent researcher at the Center for the Sociology of Innovation (CSI) of Mines ParisTech. CSI has been an extremely stimulating and friendly environment, and I thank all of its members for the support they provided. Among them, Michel Callon holds a special place. He was an irreplaceable guide in my first encounters with sociology and, later, in the development of this work. He has been a thoughtful, open-minded, and stimulating mentor, eager to help me develop new trains of thoughts.

Sheila Jasanoff welcomed me as a fellow in the Program on Science, Technology & Society (STS) of the Kennedy School of Government at the early stages of this research and a few years later, as it developed from the PhD thesis it had become. She has been a constant support and a permanent source of inspiration. I am grateful for all I have received from her, including the possibility to participate in the annual discussions of the Science and Democracy Network. These meetings have been fruitful sites for presenting and discussing many of the themes and arguments that are developed in this book.

I developed some of the ideas presented in this book from (or alongside) papers published in academic journals or collective volumes. My initial empirical explorations in Grenoble led to a paper published in *Innovation* (Brice Laurent, "Diverging Convergences: Competing Meanings of Nanotechnology and Converging Technologies in a Local Context," *Innovation: The European Journal of Social Science Research* 20, no. 4 [2007]: 343–357), which I used to develop part of chapter 6. I initiated the analysis of the political questions related to the definitions of nanomaterials in a 2013 paper (Brice Laurent, "Les espaces politiques des substances chimiques: Définir des nanomatériaux internationaux, européens et français," *Revue d'anthropologie des connaissances* 7, no. 1 [2013]: 195–221), which provided some of the first elements from which I wrote chapter 4. I presented the analysis of OECD as a site where international expertise and public engagement were problematized (as discussed in chapters 3 and 7) in a chapter written for a collective volume (Brice Laurent, "Boundary-Making in the International Organization: Public Engagement Expertise at the OECD," in Jan-Peter Voß and Richard Freeman, eds., *Knowing Governance: The Epistemic Construction of Political Order* [London: Palgrave, 2015], 217–235).

I owe a lot to the coauthors of some of the papers from which I developed sections of this book. I started a reflection on the critical engagement of the social scientist in nanotechnology with Michiel van Oudheusden, as we worked on a paper published in 2013 (Michiel van Oudheusden and Brice Laurent, "Shifting and Deepening Engagements: Experimental Normativity in Public Participation in Science and Technology," *Science, Technology & Innovation Studies* 9, no. 1 [2013]: 1–21). This was extremely helpful to work on chapter 7 of this book. I discussed the case of the French national debate on nanotechnology in a paper written with Véra Ehrenstein, in which we compared two state experiments with public participation, in France and the Democratic Republic of the Congo (Véra Ehrenstein and Brice Laurent, "State Experiments with Public Participation: French Nanotechnology, Congolese Deforestation and the Search for National Public," in *Rethinking Participation: Science, Environment and Emergent Publics*, ed. Jason Chilvers and Matthew Kearnes, 123–143 [London: Routledge, 2015]). Thinking of the French example in those experimental terms helped me develop some of the arguments presented in chapters 3 and 7 of this book. Henri Boullier and I characterized a European mode of public reasoning as a "regulatory precaution" (Henri Boullier and Brice Laurent, "La précaution réglementaire: Un mode de gouvernement européen des objets techniques," *Politique Européenne* 3, no. 49 [2015]: 30–53), which helped me analyze some of the European examples discussed in the book (chapters 4 and 7). The reflections I have been conducting with Jim Dratwa were helpful in considering responsibility as an operator that organizes the European political, economic, and moral space (as presented in chapter 5)—a perspective that I developed for this book alongside a chapter written for a collective volume published in 2016 (Brice Laurent, "Perfecting European Democracy: Science as a Problem of Technological and Political Progress," in *Perfecting Human Futures: Technology, Secularization and Eschatology*, ed. Benjamin Hurlbut and Hava Tirosh-Samuelson, 217–238 [Dordrecht: Springer, 2016]). I benefited from support from the *Transhumanist Imagina-*

tion project directed by Ben Hurlbut and Hava Tirosh-Samuelson when I worked on this latter chapter, and on the corresponding sections of this book.

As I sought to build on Science and Technology Studies to reflect on the politics of governmental action, I benefited from numerous exchanges with a group of scholars interested in technologies of participation, including Jason Chilvers, Ulrike Felt, Jan-Peter Voß, Linda Soneryd, and Sonja van der Arendt. They helped me undertake a more thorough reflection on political experiments, which directly contributed to the study of technologies of democracy presented in this book.

The editors of collective volumes who invited me to contribute to their projects have been very helpful as I developed some of the arguments presented in this book. I thank Jason Chilvers, Ben Hurlbut, Matthew Kearnes, Richard Freeman, and Jan-Peter Voß.

As friends and demanding researchers, Nicolas Benvegnu, Stève Bernardin, Benjamin Lemoine, and Liliana Doganova played a central role in my research approach for this book and helped me develop and clarify some of its arguments.

Understanding the democratic questions that nanotechnology raises required collaborating with many people directly involved in its problematization. I am grateful to the people with whom I discovered this unusual field, and particularly to the members of Vivagora and of the nanotechnology working group at the French standardization organization (AFNOR), and participants in the Working Party on Nanotechnology at the OECD. I thank the researchers of the Center for Nanotechnology in Society at Arizona State University and particularly David Guston, who welcomed me at the Center for Nanotechnology in Society (CNS) as a visiting scholar, and Erik Fisher, with whom the conversations about his projects and about nanotechnology policy in general have been fruitful.

As I extended the study of constitutional ordering to other fieldwork, I refined the questions explored in this book. As such, some of the research projects I have been involved in as the book neared completion have been important to finalize it and to start reflecting on the analytical paths it opens. These projects include the PhD thesis I have been co-supervising with Madeleine Akrich and David Pontille: Julien Merlin's study of the articulation of sovereignty, political representations and economic valuations in mining activities; and Mathieu Baudrin's ongoing work on the critical historical analysis of regulatory and market spaces related to mass consumption objects. Further informing the book are my empirical and

theoretical reflections conducted with Liliana Doganova on the joint making of democratic and market orders related to European environmental
initiatives; with Alexandre Mallard and Aurélie Tricoire on labeling practices in the construction sector; with Romain Badouard, David Pontille,
Félix Talvard and Martin Tironi on the politics of experiments in the urban
environment; and with François Thoreau on modeling and the European
regulation of chemical risks.

Prologue

A Public Debate on Nanotechnology

In February 2010, I participated in one of the sessions of the French national debate on nanotechnology. The session was devoted to the "ethics and governance" of nanotechnology. The format was quite unusual for a public debate. Because anti-nanotechnology activists had interrupted previous public meetings, the organizers had decided to adapt the process, commissioned by the French government and originally meant to be a series of public meetings opened to whoever was interested in participating. Thus, I had to fill out a form and submit it online a few days before the event. I received a response in an email in which I was asked for my mobile phone number and told to be at a Paris subway station the following day, thirty minutes before the debate was to begin. When I arrived at the designated time, a young woman handed me a map of the local area, where I could see the way from the subway station to the building where the debate would be held. After a short walk, I found the building and opened a dull and gray door with no sign on it. Two big men in dark suits greeted me and asked for my ID. Once cleared, I was shown to a staircase by which I got down to the basement of the building. At the end of a corridor with concrete walls, I finally arrived at my destination. In this closed and secret place, the debate would be protected from an unwanted public.

Like the other participants, I was sent to one of the rooms of the building to discuss in small groups issues related to the "ethics and governance" of nanotechnology. My group was quite small, and comprised, apart from myself, the president of the Commission Nationale du Débat Public (CNDP; National Commission for Public Debate) organizing the debate, a member of the French ministry of agriculture, a representative of an environmental movement partnering with industries and public bodies, and a member of a national consumer organization—all of them "friends of public debate," as the president had called them.[1] A TV crew was filming us for later broadcast on a local channel in order to "account for the fact that the debate exists."

The questions of "ethics and governance" discussed during this one-hour session were various. Some of them related to the "difficulty to locate the products in which nanomaterials were used," particularly in the food industry, where "the industry did not seem to play the game." This was problematic if the health and safety risks of nanomaterials were to be regulated, and consumers informed, as the member of the consumer organization demanded. But for the member of the ministry of agriculture, there was "no nano in food." Other interventions considered the "problem of participation," and particularly the fact that anti-nano activists "refuse to enter democratic discussions" and had forced the organizers to hold a closed debate in the first place. Eventually, CNDP's president spoke about the "ethics issue": how to construct a science policy program in an "ethical and democratic manner"? For him, this very debate was part of the answer.

As for my own role in the event, I was supposed to participate in the discussion, but felt increasingly uncomfortable. Previously, I had studied, and worked with, an association called Vivagora, which advocated the "democratization of technical choices." Vivagora had been an initial supporter of public debate, but then criticized the organizers' choice to stage closed events such as this one. I did not like the fact that my interventions could be broadcast, and my participation in this contested public debate made visible to actors like Vivagora. It was a relief that I had to leave early to meet with students. This excused me from the task of reporting the discussions, which the president of CNDP had asked me to assume.

Challenges for the Democratization of Nanotechnology

The short episode I just described is an example of "public participation in nanotechnology," and, more generally, of an objective of the French public administration to "democratize" technical choices—an ambition that is now shared by other public bodies in Europe and elsewhere. It illustrates the many difficulties this objective raises, particularly in the case of nanotechnology, and, by the same token, the challenge faced by scholars interested in the analysis of this "democratic ambition."

First, participatory devices are not external to controversies about nano-technology. The exceptional mechanism, through which the organizers had sought to exclude the opponents of the nanotechnology debate, illustrates how investments are made to shield debate from an unwanted public that conceives of participatory mechanisms not as ready-made instruments that could be "applied" to nanotechnology, but as components of a questionable nanotechnology policy. The CNDP intended the closed debate to be made public (and the TV crew was there to ensure that it would be) and used to demonstrate that "it was there" that nanotechnology was being discussed democratically. The nanotechnology debate was conceived by both its organizers and its critics as an inherent component of the development of nanotechnology. It was part of a more general ambition of the French government of opening environmental and technological policy choices to the intervention of various stakeholders. The French government was legally bound to organize a national public debate on nanotechnology because of a 2010 law—which itself originated from a nationwide consultation process on environmental legislation initiated at the beginning of the Sarkozy presidency.[2] That the organization of the national public debate on nanotechnology was delegated to the CNDP was considered an experiment in public action by members of the commission and the government itself.[3]

Second, whereas the debate was supposed to explore public concerns about nanotechnology, neither nanotechnology's problems nor its publics were clearly identified. Nanotechnology substances and products were not defined—discussions about their risks in the short episode I described evoked the question of their very existence: whether or not objects could be qualified as "nano," and therefore exposed to public scrutiny during the debate, was a stake during the discussions. During one of the meetings, representatives of the food industry claimed that they "didn't do nano," while this affirmation was challenged by public administration officials. Eventually, the definition of the topics that were to be discussed during the debate was itself the outcome of a hybrid process. The division into questions related to the health and safety risks of nanotechnology, privacy matters connected to applications in nanoelectronics, and issues related to the convergence of technologies originated from previous formalizations of anticipated issues related to nanotechnology, in American and European science policy reports (European Commission 2005; Roco and Bainbridge 2001). In the same time, the definition of the preferred discussion topics of each meeting of the national debate was done in conjunction with local industrial and research activities—hence the focus on cosmetics in Orléans, or textiles in Lille. Not only were nanotechnology's objects and problems uncertain, but so were its publics. Opponents refused the discussion, while participants to the debate were struggling to define what exactly they expected from it. Organizers would regularly complain about the fact that environmental protection organizations, consumer groups, or unions were not more numerous to intervene in the debate, and would encourage them to submit contributions even if they were not directly concerned about nanotechnology. Ironically, some of the most concerned publics of nanotechnology were precisely those that made the debate eventually fail to proceed as it was intended to: the activists who interrupted meetings had been mobilized on nanotechnology since the early 2000s, particularly in Grenoble where nanotechnology research has been a priority for public and private local initiatives.

In such a context, my external position was difficult to maintain. This is the third difficulty that the example of the French national debate on nanotechnology illustrates. As I was studying this debate as part of a scholarly initiative, I was forced to engage in a device I had first intended to observe. I eventually left, as I sensed what was not a threat to a "neutral" scholarly position, but a conflict between speaking publicly within this device and my engagement with the actors I was studying. Eventually, the intervention I was engaged in challenged the distance between scholarly description and normative intervention: as nanotechnology is an entity in the making, and made as much by devices such as the national public debate that contributes to stabilize its boundaries, the public problems it raises, and the identity of its publics, the analytic intervention is bound to participate in the making of the reality it seeks to account for.

Analyzing Democracy and Nanotechnology

This book studies the democratic constructions entailed by the definition and public treatment of nanotechnology problems. It argues that nanotechnology, precisely because of the characteristics introduced previously, is a lens through which one can develop a theoretical and practical approach to the study of contemporary democracies. It explores the joint production of nanotechnology itself and democratic order and identifies questions related to the exercise of citizenship, the forms of national sovereignty, and the channels of political legitimacy within the very making of nanotechnology as a heterogeneous entity comprising objects, futures, concerns, and publics.

The approach I follow is comparative. Taking inspirations from recent works in science and technology studies (STS) and political theory, I use comparison as a way of identifying the contingency of technical and political choices, and the imbrications of democratic constructions with national or supranational institutional arrangements. The cases I consider are related to national and international contexts. I discuss empirical examples from the United States and France on the one hand, from European institutions and international organizations on the other hand. Nanotechnology originated as a science policy program in the United States, where the impetus for international competition and collaboration was initiated. By contrast, France defined its own national nanotechnology strategy relatively late. Yet France has its own specificity: it is the only place where anti-nanotechnology protests have adopted such a radical mode of critical intervention. France also holds a specific position at the European level, by arguing for constraining regulatory choices and being the first country to introduce, in 2011, a mandatory declaration of "substances in a nanoparticulate state." These two national cases are directly connected to international arenas: while the definition of national choices in Europe is tightly connected to the European regulation, countries involved in the development of nanotechnology are active in the standardization of the field in international organizations such as the International Standardization Organization (ISO) or the Organisation for Economic Co-operation and Development (OECD).

Through the examination of the constitution of nanotechnology as a collection of objects, futures, concerns, and publics, this book seeks to develop a theoretical and practical approach for the study of democracy as an entity in the making, potentially contested, and challenged in a variety of empirical sites. In framing nanotechnology as a large-scale science policy program covering a wide range of scientific disciplines and practices, nanotechnology's proponents were eager to take democratic issues into account. This makes nanotechnology a relevant focal point for a renewed analysis of democracy. Throughout the subsequent chapters, I argue that STS has much to offer in order to renew democratic theory. Accordingly, this book is situated within an interest for political ordering that stresses the central role of science and technology in the making of the democratic life, the imbrications of scientific projects and programs with nation building, and the importance of national political culture for the enactment of particular modes of objectivity and legitimacy building (Jasanoff 2005, 2012). In addition, it builds on a series of work in the post-Actor–Network Theory vein that has accounted for the distribution of agency in the stabilization of political and economic organizations, particularly by using notions such as "agencements" (Callon 2007). In this book, I follow these trends of work in order to reflect on the ways in which STS can renew democratic theory. I argue that, far more than a simple call for "public participation in science and technology," an STS-inspired democratic theory articulates three components.

First, it redefines the extent of democratic issues and relocates the places where democracy is at stake. The following chapters focus on the sites where nanotechnology is defined as a public problem. They analyze the *problematization* of nanotechnology, that is, the ways in which nanotechnology is defined as a problem worthy of collective examination and treatment. As I will argue throughout the book, sites of problematization are the places to examine in detail in order to account for democratic ordering. These sites comprise, but are not limited to, participatory mechanisms such as the French public debate on nanotechnology. This is indeed one of the core arguments of this book, that grasping the ongoing evolution of contemporary democracy—such as calls for extended public participation or the anticipation of public problems—requires an analysis able to connect various sites of problematization, from science museums to regulatory institutions, from public debates to secluded standardization organizations. This book argues that it is by examining sites often overlooked by political theorists, and indeed considered at the margins of democratic life, that one can develop richer accounts of contemporary democracies. Technological development suggests rethinking democratic theory from the margins of political institutions because it is from these margins that democratic ordering processes are explicitly questioned.

Rethinking democratic theory in these terms requires that one accounts for the destabilization and restabilization of institutional constructs. This is a second component of the perspective on democratic theory proposed in this book. It argues that problematizing nanotechnology is also problematizing democracy, including the exercise of citizenship, the crafting of legitimate decision-making pathways, and the definition of sovereign public action. It does so through the analysis of different types of instrumented public initiatives where the nature of democracy is put to test. The validity of these "democratic experiments" depends on their imbrications within larger institutions that might be national or supranational, and that might be restabilized or displaced because of these experiments. Thus, understanding the problematization of nanotechnology requires an analysis of institutions such as public expertise bodies, regulatory organizations, and parliaments, conducted from the sites where the rules organizing democratic life are questioned.

Third, the approach proposed in this book redefines the intervention of the analyst in ways that do not fall into such dualisms as "descriptive" vs. "normative" or "neutral" vs. "engaged." In the following developments, I refrain from qualifying from the start what is democratic and what is not, and maintain a certain agnosticism about democracy. I voluntarily start the analysis with an open definition of "democracy," based on the definition and treatment of public problems (this will be discussed in chapter 1). As seen in the opening scene, my own engagement was at stake in the conduct of this study: the approach proposed in this book will develop a perspective on the modalities of scholarly engagement. The objective is not to reproduce well-known categories, such as "participation observation," which, I will argue, fail to account for the multiple forms of engagement needed to both conduct empirical work on such elusive entity as nanotechnology and develop a consistent critical approach on democratic ordering. By considering that the intervention of the researcher is part of problematization processes, this book proposes to rethink the normative objective of democratization by insisting on the stabilization and destabilization processes of problematization.

The following chapters develop this perspective on democratic theory by focusing on sites of problematization of nanotechnology: problematizing nanotechnology, as I argue throughout this book, is both constituting it as a heterogeneous entity and organizing democratic order. Accounting for the construction of nanotechnology and democracy implies a methodological and theoretical reflection on the methods for an analysis of the democratic problems raised by nanotechnology. The challenge is to grasp "nanotechnology" as a hybrid, contested, and fluid entity, to locate the site where it is problematized, and to understand in what ways nanotechnology raises issues for the democratic organization. This objective will be clarified in the first chapter, where I discuss the methodological approach that I follow in this book. I then turn to the examination of sites where nanotechnology is problematized: sites, like science museums and participatory devices, where it is represented for publics (chapters 2 and 3); sites in national administration and international bodies where categories for nanotechnology objects are crafted (chapter 4), and conditions for the "responsible" development of the field are defined (chapter 5); and sites where more or less organized social movements engage in its external or internal critique (chapter 6). The successive chapters thereby propose to reexamine well-known operations of democratic life from the sites where technological development is problematized, namely the representation of objects and publics, the government of material and human entities, and the engagement of various social groups, including that of social scientists themselves. Eventually, they will allow me to introduce a critical perspective on the democratization of technology, which I label "critical constitutionalism."

1 Problematizing Nanotechnology, Problematizing Democracy

A Political Entity

Objects

In January 2007, I met with Patrick Boisseau in his Grenoble office at a laboratory of the Commissariat à l'Energie Atomique (CEA) called LETI.[1] Boisseau, a biologist for CEA since 1987, had become the coordinator of the "European network of excellence" known as Nano2Life that was funded under the European Commission's Sixth Framework Programme for Research and Technological Development. Nano2Life gathered twenty-three research institutions in ten different countries across Europe. As a "network of excellence," Nano2Life did not add new research projects to those conducted by the partners. Rather, its purpose was to "reduce fragmentation in European nanobiotech" by undertaking various common initiatives, such as training programs in nanotechnology, circulation of research staff among partners, sharing of scientific equipment, and coordination of long-term research objectives among the partners.

The partners of Nano2Life would share their research methods, confront their results, and attempt to align their projects involving physicists and biologists. The range of cooperation between different disciplines was, for Boisseau, quite a new phenomenon. He told me that "the idea was really to bring together physics and biology, and use both of them for the development of new devices." By this he meant nanoparticles (that is, particles composed of fewer than 1,000 atoms) that could be used as tracking devices inside the human body for imaging, or as drug delivery devices (called "nanovectors"), bringing the drug to the very cell in need of it. "Regenerative medicine" was also a topic of inquiry, since "smart biomaterials" could be developed, that is, small-size components precisely targeted to be added to a human tissue. Nanoscale-diameter fiber implants ("nanowires") could conduct an electric stimulation to a precise location in the body, for instance, in the brain—the long-term objective being nothing less than to cure Parkinson's and Alzheimer's diseases.[2] Thus, Nano2Life was meant to bring together laboratories working with "nanoscale objects," designed to offer new properties thanks to the small size of their components. The laboratories involved in Nano2Life produced numerous objects made of assemblages of metallic atoms and biological molecules, implants and wires, nanoparticles and nanocoatings. Boisseau was enthusiastic about what the nanoscale could bring: nanoparticles could bear completely different chemical and physical properties from their non-nano counterparts, and, associated with biological materials, could pave the way for a "new biomedicine," tailored to the exact needs of the patient.

The objects Nano2Life participants produced have an uncertain status. Nanoparticles are "new" substances in that they provide new properties (thanks to which, for instance, metallic particles can be used as tracking devices inside the human body, or to carry molecules of drugs). But how they differ from their non-nano counterparts, and whether, for instance, they are considered as "new particles" in current regulation is unclear.[3] Within the European legislation, medical objects are regulated as either "products" or "devices," the former requiring stricter regulation (and constraining rules about human testing) than the latter. But whether a nano-vector is a product or a device is hard to tell. This could be problematic, since these objects, developed for medical applications, would require human testing to be finalized. The uncertainty about where the Nano2Life objects fall in the regulatory landscape is not a detail. It is a sign of the transformation nanomedicine proposes to bring to the conduct of scientific research, bringing together both physics and biology, applied medicine and fundamental research, human testing and upstream research, while paving the way for a medical discipline that attempts to specify its interventions according to the individual needs of the human patient, and, even more, to the needs of the patient's each individual cell.[4]

Nano2Life's material productions "have politics," to use Langdon Winner's famous phrase, in that they inscribe users and long-term objectives in the organizations of health care (Winner 1980).[5] More generally, and taking into account the flexibility of these objects themselves, one could interrogate the transformations they propose. How far do they challenge legislation, industrial strategies, the conduct of clinical trials, and the status of experiments with humans? Answering these questions is exploring democratic issues related to the entry of new material elements in society. To use a Latourian vocabulary, these objects reconfigure heterogeneous associations and make new ones emerge.[6] They are material elements to take into account in the construction of a common world. They could be more or less equally distributed. They could benefit private companies, or be openly shared. They can offer new routes for the conduct of medical research based on the rapid development of applications, close relationships between physicists and biologists, and the blurring of boundaries between laboratory experiments and the development of medical treatments.[7]

Uncertain Nanotechnology

The material dimension of nanotechnology is problematic, however. For one can wonder what makes Nano2Life objects "nano." If "nano" points to the manipulation of matter at the atomic level, then it is best understood in terms of its scientific instrumentation—the main representative of which being the scanning tunneling microscope (STM), which, by using the quantum "tunnel effect," can picture individual atoms while simultaneously moving them. It made it possible to manipulate matter "atom by atom"— an idea that was central in the successive books of a scientist turned futurist, Erik Drexler, who advocated the development of "molecular manufacturing," by which "nanomachines" would be sophisticated enough to reproduce themselves.[8] What constitutes nanotechnology was then the topic of lengthy debates. Drexler and famous nanotechnologist Richard Smalley, Nobel Prize winner in chemistry, the discoverer of fullerenes, and a key proponent of U.S. nanotechnology programs, opposed each other in the early 2000s in a series of articles about the feasibility of molecular manufacturing. The opposition can be summed up by what philosopher of science Bernadette Bensaude-Vincent called the "two cultures" of nanotechnology: while Drexler imagined using mechanical methods to manipulate atoms and construct nanomachines, Smalley, a proponent of a chemistry-based approach, contended that the mechanical "fingers" would be too "sticky" to manipulate atoms (Bensaude-Vincent 2004). The opposition was not limited to academic circles. When the U.S. National Nanotechnology Initiative (NNI) was constituted in the late 1990s, Drexler argued that the NNI had sold nanotechnology to business interests, while representatives of private companies considered Drexler's visions as little more than "a wino's claims."[9] The former considered that the NNI had gone "from Feynman to funding," that is, from a grand and path-breaking vision prophesized by Nobel Prize winner Richard Feynman in the late 1950s and made of self-replicating nanomachines, to a collection of disparate projects, only gathered together because of their use of small-size objects, and, above all, their economic prospects (Drexler 2004). The latter contended that Drexler's arguments were little more than science fiction, at best unrepresentative of what nanotechnology was in the concrete functioning of laboratories and businesses, at worst threatening to the general public, who could become skeptical of nanotechnology if fed with too many stories of self-replicating nanomachines potentially escaping human control—a hypothetical risk Drexler himself had discussed in his work.[10]

Hence the quality of being "nano," the "nano-ness" of objects and programs, is not uncontroversial. Nano2Life gathered a number of objects, some already existing, others foreseen in the future, some based on isolated chemicals and others more sophisticated. Nano2Life was not concerned with molecular manufacturing as Drexler imagined it, but it did propose to use biological structure to construct molecular machines. Nano2Life also included in its objectives the development of nanoparticles that had been known for years, and which were, thanks to their integration in the project, rebranded as "nano." What made Nano2Life objects "nano" was—more than a single definition based on a scientific process (as genetic engineering could define biotechnology) or a material technology (as the computer could define information technology)—their integration in science policy programs expected to attract public attention and support and based on technical interventions at the atomic scale.

Consequently, looking at nanotechnology objects raises a fundamental difficulty: does the analyst need to distinguish between "true" and "false" nanotechnology, as authors trying to decipher the "nano hype" in order to identify "the truth behind it" would lead us to think (Berube 2006)? As I will argue throughout the book, the contested and uncertain qualification of "nano" is to be the main focus of analysis if one wants to grasp the democratic challenges of nanotechnology. This requires considering that what makes an object nano is not considered as given, but as the outcome of negotiations among actors, involving the evaluation of new physico-chemical properties, strategic economic considerations, and the construction of science policy narratives and instruments.

Programming Nanotechnology
As a "network of excellence" of the EC's Sixth Framework Programme—the only one in nanobiotechnology, Nano2Life was a central component of the European research policy. It was expected to "reduce the fragmentation of European research," and, as such, part of a series of initiatives aiming to organize the European Research Area according to the long-term objective of the Lisbon strategy, namely the "transformation process (of Europe) into a knowledge-based economy."[11] Nano2Life was expected to answer a growing concern within European science policy circles: that

European nanotechnology research was lagging behind that of other developed countries, most notably the United States, within what had become a "global nanorace" (Hullmann 2006a). The race for public funding went with a race for promises. The objectives of Nano2Life (such as "revolutionize cancer treatment") appear almost moderate when compared with the promises made by the proponents of U.S. nanotechnology programs, which presented nanotechnology as no less than the "next industrial revolution" (McCray 2005). But they were also situated within the development of a science policy program, expected to ensure the integration of the European Research Area as well as stimulate technological innovation.

As it appears through the example of Nano2Life, nanotechnology can be described as a science policy program, which integrates research projects for explicit, long-term strategic objectives supposed to be relevant for collective action. It is discussed in public institutions (such as the European Parliament or the American Congress), and administrative bodies. It is tied to questions of economic dominance of countries or international political spaces (such as the European Union). Indeed, as the measure of performance in the "global nanorace" requires common definitions of what is "nano" and what is not, nanotechnology is also an object of international concern, in that norms and standards were called for early in the development of nanotechnology programs.[12] Are nanotechnology objects and the long-term objectives presented in statistics and promises disconnected? This is what some commentators might lead us to think while trying to identify "the truth behind the hype," that is, the "real" laboratory practices that would lie "behind" the grandiloquent policy discourses based on futuristic promises and sustained by competitiveness arguments. But if one does not accept the dichotomy between "real" scientific practices and "false" nanotechnology, understanding the connection between the objects such as those Nano2Life's laboratories produce and the long-term objectives forces one to consider another component of nanotechnology: that of the instruments of science policy.

Futures

Patrick Boisseau was the coordinator of Nano2Life. He was also involved in many local, national, and European projects. He had supervised the organization of the "Nanobio innovation center" in Grenoble, a research center sponsored by CEA and the local university, which was meant to develop "new miniaturized tools for biological applications." Like others in Europe and the United States, Boisseau was a scientist who had become actively involved in the management of nanotechnology research. And like many other proponents of nanotechnology, he considered that his intervention had to target both technological developments and the social and political organization that were supposed to make them happen.[13] Boisseau's involvement in the local organization of nanotechnology research was tightly linked with the construction of the European nanotechnology policy. During our discussion, Boisseau gave me a "vision paper" about nanomedicine, which had been released in 2005. He had participated in the writing of this publication for the European Union, which was the first step in establishing a "European Technology Platform" for nanomedicine ("ETP nanomedicine").[14]

The vision paper to which Boisseau contributed was crafted along the same themes as Nano2Life. It emphasized "Nanotechnology-based Diagnostics including Imaging," "Targeted Drug Delivery and Release," and "Regenerative Medicine," each of them illustrated by examples, such as "nanoanalytical tools" "incorporated into 'lab-on-a-chip' devices, which can mix, process and separate fluids, realizing sample analysis and identification" (European Technology Platform on Nanomedicine 2005, 16), "microfabricated device with the ability to store and release multiple chemical substances on demand" (ibid., 24), or "'intelligent' biomaterials (...) designed to react to changes in the immediate environment and to stimulate specific cellular responses at the molecular level" (ibid., 28). For each of the three categories of products, the vision paper proposed the "basis for a strategic research agenda." For instance, with regard to nanoprobes, the vision paper stated that biocompatibility was to be improved.

The vision paper was only a preliminary step before the construction of a roadmap for European nanomedicine, written by researchers, industrialists, and officials from the European Commission, and to which Nano2Life directly contributed. Part of Nano2Life's activities was the organization of "foresight exercises," through which the project could "identify the future applications or techniques to focus the research efforts on."[15] The roadmap that emerged from the ETP nanomedicine and Nano2Life was meant to coordinate European research and define objectives for the next nanotechnology policy initiatives. It identified problems to be solved and potential outcomes. For instance, it defined "devices for drug delivery" as "targeted applications," and then pointed to "key R&D priorities" (e.g., biocompatibility of materials and miniaturized systems), needed technologies (e.g., "nanocapsules"), the "challenges" to be met (e.g., the stability of the device), and diseases supposed to be cured (cancer, diabetes, or cardiovascular disease) (European Technology Platform on Nanomedicine 2005, 30). The roadmap considered that nanotechnology required the early identification of promising domains and the definition of appropriate research funding flows. Fundamental and applied research had to come together, and the roadmap heralded "public-private partnerships" as instruments through which nanotechnology could be developed according to the objectives defined, with limited public funding support.

From the example of Nano2Life, nanotechnology appears as the outcome of science policy initiatives that connect developments in laboratories and long-term perspectives, material productions of objects, scientific results, and expectations about the future. The roadmap that originated from Nano2Life and the ETP nanomedicine proposed a construction of nanobiotechnology bringing together administrative, industrial, and scientific actors in the making of a technological domain connecting industry and academic research, fundamental and applied research, for the sake of the economic and social European development. Initiatives such as this roadmap thereby pursue a trend that originated in materials science (Bensaude-Vincent 2001), and take it to yet another level, that of global funding plans for research. The roadmap was not meant as a representation of nanotechnology that could have been assessed according to the accuracy of its description of a given scientific reality, but it actively contributed to produce nanotechnology by gathering scientists, rationalizing current developments in scientific laboratories, reflecting on their potential evolutions, and eventually operationalizing them in the making of European nanotechnology programs. Nano2Life's motto was "bringing nanotechnology to life": it was as much about applying nanotechnology to biological applications as about making nanotechnology exist.

Nano2Life was not the only component of nanotechnology-related policy initiatives in Europe. Other programs in materials science, electronics, and environmental sciences were launched, within a global European nanotechnology strategy, presented in the European Commission's report *Nanosciences and Nanotechnologies: An Action Plan for Europe 2005–2009*, released in 2005 (European Commission 2005, 243). The *Action Plan* aimed to make the European research area a major actor in nanotechnology research. This required, as seen in the example of Nano2Life, the operationalization of the future of nanotechnology in science policy instruments. In the United States as well, the future of nanotechnology was operationalized in roadmaps and programs of development. The most visible of these instruments is certainly the "four generations of nanomaterials," presented in a graph made by Mihail Roco, the director of the U.S. National

Nanotechnology Initiative (NNI). It proposed a synthetic vision of the development of nanotechnology, in which "passive nanostructures" were followed by "active nanostructures," "systems of nanosystems," and "molecular nanosystems." When Roco published the graph in 2004, the last three generations were to be developed in the future (figure 1.1).

The four-generation graph circulated widely. Roco displayed it at numerous academic conferences and workshops. It was presented in numerous science policy circles (such as the meetings of the ETP nanomedicine). It was used as a reference by science policy officials outside of the United States. For instance, the official in charge of the nanotechnology research programs at the French national agency for research (ANR) explained the organization of the funding plans for nanotechnology research by making direct reference to Roco's graph. "The domain we need to explore," he explained to me during an interview, "is the development of nanosystems," which was "Roco's fourth generation, the final step."[16]

The organization of research defined as such directs funding flows and stimulates particular trends of technological development. It is based on the constitution of networks among laboratories sharing knowledge

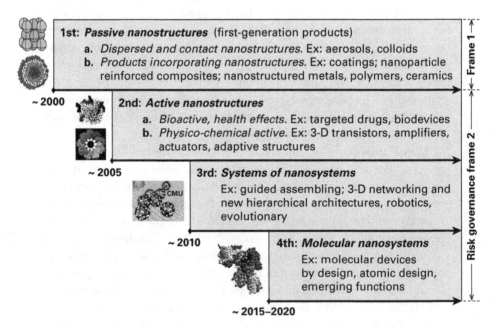

Figure 1.1
Four generations of nanomaterials (Roco 2004)

and infrastructures.[17] It also aims to recompose the boundaries between fundamental and applied research, and among scientific disciplines.[18] The long-term objectives (economic competitiveness, transformation of research/industry collaboration, development of new medical tools) of nanotechnology programs are directly connected to the material construction of objects through the instruments of science policy. As Miller and O'Leary put it, these instruments "link science and the economy through acting on capital budgeting decisions, and in doing so (...) they contribute to the process of making markets" (Miller and O'Leary 2007, 702), as well as, one could add, laboratory practices and public decision-making processes.[19]

The instruments that make the future of nanotechnology—funding plans, roadmaps, and science policy programs—operationalize conscious public choices, such as developing technologies for economic competitiveness, answering "social goals" (e.g., curing diseases), and establishing long-term R&D objectives (such as molecular manufacturing or molecular nanosystems).[20] This implies that nanotechnology is the outcome of collective decisions to be made about the future (e.g., allocating public money for nanotechnology research, and developing particular technological areas rather than others), which involve scientists, industrial actors, and public officials in hybrid arenas (like the nanomedicine ETP): this means nanotechnology's future is a topic for collective decisions, and an issue for democratic societies.

Concerns

As we discussed Nano2Life, Boisseau immediately mentioned the network's "strong concern for ethical issues." An "Ethical, Legal and Social Aspects (ELSA) board" had been set up since the beginning of the project—and Boisseau proudly gave to me one of its publications (Ach and Siep 2006). The ELSA board was the first attempt at institutionalizing ethical reflections in Europe in the field of nanobiotechnology. It followed the requirements of the European *Action Plan*: that nanotechnology's ELSA should be taken into account in European research. The vision paper of the ETP nanomedicine devoted a section to "regulatory issues and risk assessment," and another one to "ethical issues." For Boisseau, the concerns with nanotechnology were indeed either "related to risk" or "ethical ones." The former dealt with the potential adverse effects of nanoparticles and nanomaterials for human health and the environment. Nanotechnology produces substances bearing enhanced properties that—as Boisseau was aware—could also have toxicological reactivity that differed from their non-nano counterparts. Ethical concerns were much vaguer in Boisseau's discourse: they referred to "problems of informed consent," "issues of fair repartition of benefits," and "long-term issues"—by which he meant philosophical questions related to the use of biological materials for the making of (still hypothetical for most of them) nanomachines, and issues related to the use of nanotechnology for "human enhancement."

That nanotechnology could raise public concerns is not surprising. By bringing new objects into life, scientific research is bound to do so. Biotechnology, for instance, produces new living organisms by genetic manipulation, transforms embryos into research objects, turns living material into patentable goods, and, eventually makes "life itself" a public concern (Rose 2001; cf. Jasanoff 2005; Rajan 2006). Nanotechnology does not seem to be different from other domains of scientific activities, for that matter. This was well recognized by Boisseau, as he explained that Nano2Life had felt compelled to set up an ELSA board because of the "questions nanotechnology raised, as any other technology." But as opposed to stem cells, embryos, genetically modified organisms (GMOs), or nuclear waste, nanotechnology "objects" are not easily identifiable. They gather medical products, chemical substances, commercial products, laboratory objects, and future developments that exist nowhere but in roadmaps and strategic plans. For Boisseau, the previous experience of past controversies and the fact that nanotechnology objects were still in the development phase forced scientists to "make it right." He meant that Nano2Life, and, even more, European nanotechnology policy had the "obligation" to identify and deal with nanotechnology concerns even before problems or controversies emerged. In previous cases indeed, such as biotechnology, questions related to the risks of objects as GMOs, or ethical issues related to "messing with nature," have caused considerable controversies, whether related to risk evaluation or to the reduction of ethical questions concerning risk evaluation.[21]

Boisseau's call to "make it right" was not an isolated proposition. When Mihail Roco and William Bainbridge, the two people in charge of the American National Nanotechnology Initiative, organized in the early 2000s a series of meetings about the "societal implications of nanotechnology," they explicitly took in charge the concerns that nanotechnology might raise (Roco and Bainbridge 2001, 2003a, 2005). For them, for nanotechnology to be a success it was necessary to integrate the study of these "implications" in the very making of programs—by that they meant that the potential safety risks of nanomaterials were to be evaluated and taken care of at an early stage, and that the potential ethical issues of nano objects (such as the informed consent of patients involved in medical trials, or the question of "human enhancement" through nano devices) were addressed.

When the U.S. 21st Century Nanotechnology Research and Development Act of 2003[22] was passed, it required the "integration" of the study of "social impacts" within nanotechnology federal programs. The European Commission's *Action Plan* also called for the study of the "ethical, legal and social aspects" of nanotechnology (as noted earlier, known as ELSA), which prompted the funding of numerous European projects meant to answer questions such as these: "What will society look like when nanotechnology becomes more mainstream? Will the products be profitable? Are there any negative environmental or health impacts? Who controls the use of nanotechnology? How to deal with liability? Whom will the technology benefit or harm? What are the ethical problems?" (Hullmann 2006b, 7).

The last question clearly shows that the "ethical problems" of nanotechnology were then far from determined. On the contrary, the ELSA projects were expected to anticipate their emergence by exploring as early as possible the potential issues they could raise.

The "integration" of ELSA in nanotechnology programs means that research projects related to "implications" are funded as part of nanotechnology programs, some involving social scientists, others led by toxicologists or environmental scientists. In Nano2Life, ethicists and scientists were supposed to work closely together. How the interest in nanotechnology's "social impacts" or "ELSA" relates to the actual making of nanotechnology products and applications, and to the construction of nanotechnology's future, is a question to ask. It will be done in the following chapters. At this stage, it suffices to consider that nanotechnology is composed not only of objects and futures, but also of "concerns." Rather than taking these concerns for granted, I consider that they are part and parcel of the science policy programs that make nanotechnology take shape. They are outcomes of collective actions that result in the definition of problems considered legitimate and worthy of examination. There is at this point a democratic issue: how are these collective concerns defined, and for whose interests? Are there alternative definitions of what the relevant concerns are?

Publics

John Dewey famously argued that in a democratic society, "publics" emerge when problems are not adequately dealt with in existing institutions (Dewey [1927] 1991). Whether or not Patrick Boisseau had read Dewey, the mechanism of the emergence of publics he outlined to me was not far from the pragmatist understanding of publics and problems: "Well, it's as simple as that. If there is trouble, if there is a health crisis, then the public will not accept this. It is crucial not to do the same thing as GMOs. I think it's something everybody is aware of."

By this he meant that the European public had rejected GMOs—an argument routinely used by nanotechnology proponents in administrative circles,[23] which caused his cautious attention to the "public of nanotechnology." Nanotechnology's public was yet another component of the Nano2Life project, which included in its objectives the "education of society" and the "dialogue with civil society." The former related to training programs for students, and materials aimed to communicate the outcomes and objectives of Nano2Life. The latter pointed to the identification of public concerns—for instance, thanks to the ethics board. The other components of the European nanotechnology policy, in heralding the "societal dimension" of nanotechnology, also insisted on the need "to establish an effective dialogue with all stakeholders, informing about progress and expected benefits, and taking into account expectations and concerns (both real and perceived) so to steer developments on a path that avoids negative societal impact" (European Commission 2005, 8).[24]

The call for "public dialogue" and the consideration of "citizens' expectations and concerns" (Hullmannn 2006b, 12). was not limited to Europe.[25] Following the reports released by the U.S. National Science Foundation about the "societal implications of nanotechnology," in which the need for "two-way communication with the public" had been expressed (Roco and Bainbridge 2001), the U.S. Nanotechnology Act required that U.S. nanotechnology programs

ensure that ethical, legal, environmental, and other appropriate societal concerns, including the potential use of nanotechnology in enhancing human intelligence and in developing artificial intelligence which exceeds human capacity, are considered during the development of nanotechnology by providing (...) for public input and outreach to be integrated into the Program by the convening of regular and ongoing public discussions, through mechanisms such as citizens' panels, consensus conferences, and educational events, as appropriate.[26]

When integrated in nanotechnology policy, the mobilization of publics becomes part of what is to be discussed and decided about nanotechnology. It requires instruments expected to represent nanotechnology for "the public," to "inform about progress and expected benefits" (to re-use the language of the European Commission's *Action Plan*), and also devices aiming to "take into account expectations and concerns."

For Boisseau, the interest in nanotechnology's publics had a vivid significance. He had been confronted in Grenoble with anti-nanotechnology groups that had transformed the peaceful French Alps town into the scene of violent oppositions against nanotechnology (Laurent 2007).[27] Boisseau had then participated in public meetings about nanotechnology, sponsored by the local elected bodies as a response to this opposition. When I met him, he was skeptical about these meetings: "People did not really touch on the real problems," he said to me. He went on: "In Nano2Life, the ethics board managed to do far better, and provided concrete outcomes that will then be brought back to the Commission." He meant that Nano2Life had discussed at length the issue of the fair repartition of nanotechnology benefits, and was a cautious voice on "human enhancement" through the use of nanodevices in the human body—which was indeed restated in other publications of the European Commission about nanotechnology. The Nano2Life ethics board was, for Boisseau, a channel via which to represent the public and, by the same token, to represent nanotechnology objects, futures, and concerns for the public to understand them. This was not the only way of conceiving the production of nanotechnology publics: the anti-nanotechnology activists marching on the streets of Grenoble offered a clear contrast.

Like Patrick Boisseau in Grenoble, nanotechnology actors (whether public officials, scientists, or, like Boisseau, mediators between the two) struggle with the dialogue mechanisms to organize, the actors to talk to, and the interventions of critical groups. The following chapters examine how to set up devices meant to "make the public speak" and connect them to the production of concerned groups seeking to intervene in the making of nanotechnology. At this stage, one can see that nanotechnology is as much about publics as it is about objects, futures, and concerns. For the analysis of nanotechnology, this means that the interesting question is not about the "true" representation of public opinion about nanotechnology, but about the instruments that are used to manufacture the publics that are supportive of or involved in the making of nanotechnology, and contribute to stabilize it.[28]

Nanotechnology as a Heterogeneous Entity

The example of Nano2Life shows that nanotechnology is a broad entity that gathers material substances and products constructed in laboratories, promises of future realizations, definitions of public concerns, and publics with roles to specify. Each of these components raises issues for the democratic organization: how to integrate new material elements in society?

How to collectively define future developments of technology? What are the legitimate public concerns to deal with? How to represent publics expected to voice their "expectations and concerns"? An articulation of objects, futures, concerns, and publics such as Nano2Life can be seen as a proposition to answer these questions. It then follows that the analysis of nanotechnology making is also an examination of the challenges for the democratic organizations. Reciprocally, studying the democratic issues related to nanotechnology implies examining the making of nanotechnology as an articulation of objects, futures, concerns, and publics.

This makes nanotechnology a particularly interesting case for a reflection on contemporary democratic life. As the components of nanotechnology raise questions pertaining to the functioning of contemporary democracies, they constitute trials in which the meaning and consequences of democracy have to be reexamined. This implies that the analytical question, for anyone wishing to explore the democratic issues raised by nanotechnology, is not that of nanotechnology's "implications" for democracy, but that of the ways and means of the mutual constitution of nanotechnology and democracy itself. It is then necessary to not take for granted the discourse of nanotechnology promoters concerned about nanotechnology's "implications" and their "governance." Consider, for instance, the ways in which Ortwin Renn, a well-known specialist of risk perception studies, and Mihail Roco, the director of the NNI, argue that nanotechnology is in need of "a switch from government alone to governance." They consider that instead of "a top-down legislative approach which attempts to regulate the behavior of people and institutions in quite detailed and compartmentalized ways," a system in which "people and institutions behave so that self-regulation achieves the desired outcomes" has to be put in place (Renn and Roco 2006).[29] In Renn and Roco's perspective, the governance system should include the examination of the health and safety risks of nanomaterials at an early stage in the development of nanotechnology products; international initiatives to promote common standards able to ensure the safety of nanotechnology objects; and permanent interrogation of nanotechnology's existing and future ethical issues through the mobilization of social scientists as well as dialogues with "the public." This requires a "coordinated approach" comprising the standardization of products, training programs for scientists and social scientists, regular assessments of public perceptions of nanotechnology, and careful risk examination. Renn and Roco's approach is a synthetic version of nanotechnology programs as developed in Europe as well as in the United States. The "governance system" they propose cannot be separated from nanotechnology itself: it is a condition for nanotechnology to exist. For anyone wishing to understand the making of nanotechnology and its stabilization within democratic societies, Renn and Roco's approach is not a ready-made solution to follow but should instead be considered as a phenomenon to analyze: How does such a "governance system" come to be stabilized? How does it translate in the transformation of nanotechnology into a concern for democracies? What form of democratic organization does it enact?

This means that the study of nanotechnology implies the joint examination of the construction of scientific knowledge and technical objects, and of the production of public management approaches for new objects, decision-making processes on future developments, definitions and treatments of collective concerns, and forms of representation and mobilization of publics. Studying nanotechnology in these terms means exploring the "coproduction" of technical and democratic orders. This term introduced by Sheila Jasanoff (Jasanoff 2004) is useful here to point to the fact that as nanotechnology is being crafted, then democracy is also at stake.

Democracy and Problematization

Nano2Life raises various democratic questions. Their common characteristic is that they are related to problems to solve and decisions to make, about the objects, futures, concerns, and publics of nanotechnology. In the following chapters, I hypothesize that democracy is at stake in the places where public problems are made explicit and potential solutions are publicly explored and selected. This is a minimal definition that considers democracy as a category in the making and is not intended to be operationalized in criteria that could discriminate what is "democratic" and what is not. Rather, it is meant to help me point to the sites where nanotechnology and the democratic order are coproduced. This minimal definition echoes theoretical reflections on democracy as a political format organizing "the healthy and overt expression of conflicts of interest and differences of judgment" and defining processes for "choices to be made, opinions to be selected, and conflicting interests to be reconciled" (Rosanvallon 2011a, 119). Following a path opened by political theorist Claude Lefort, this definition considers democracy as the political form that both institutionalizes opposition and ensures the indeterminacy of the evolutions of collective life (Lefort 1986, 25–30).[30] Rosanvallon's or Lefort's approaches to democratic theory might propose general criteria according to which one could identify what is democratic and what is not, but which might not be easily specified (as epithets such as "healthy" and "overt" in Rosanvallon's quote earlier suggest). I do not attempt to engage such reflection at this point. Rather, I use these insights in order to point to particular empirical sites where the nature and modalities of democratic life are engaged. In doing so, I am also taking inspiration from historian Pierre Rosanvallon's proposition to "start from the *problems* democracy must resolve" in order to "investigate different national or historical experiences" (Rosanvallon 2008, 26). The "problems" Rosanvallon is interested in are those of political philosophy, for instance, "the tension between the sociological and the political principles of representation" (ibid.). By contrast, analyzing technological developments such as those related to nanotechnology invites us to start from practical problems raised as these developments occur, and develop from there an analytical approach through which these problems can be accounted for and engaged with. Accordingly, I will focus on empirical sites where nanotechnology is *problematized*, where the construction and articulation of its objects, futures, concerns, and publics are made a collective problem, and for which solutions (be they technical, procedural, institutional, or related to social mobilization) are crafted. Eventually, the challenge is to explore the problematization of nanotechnology as a lens for the study of the problematization of democracy itself.

I will follow a comparative approach in the subsequent chapters, to account for the variety of the problematizations of nanotechnology, and, by the same token, for the variety of democratic constructions. The example of Nano2Life is European by nature. It is also connected to French initiatives and protests, to the American early initiatives in the development of nanotechnology policy, and to international standardization projects. The following chapters will discuss American, French, European, and international sites of nanotechnology problematization, and thereby identify various democratic constructions. The challenge, then, is both to identify the sites where nanotechnology is problematized, and to describe the democratic orders that emerge out of them. It is only at a later point that a reflection on the critical approach to democratization will be possible.

Problematizations of Nanotechnology

Problematization and Political Science
How can one explore the problematization of nanotechnology? The use of the term "problematization" that I propose here stems from various bodies of work, particularly Foucault's later works and Science and Technology Studies (STS). But there is also a body of literature in political science that

focuses on public problems and the ways of dealing with them, and does not consider as a given their collective dimension. Accounting for the joint problematization of nanotechnology and democracy requires making some differences explicit, though. Consider, for instance, agenda studies, which analyze the mechanisms through which a problem is included in the functioning of political institutions, these mechanisms being determined by a series of social variables (e.g., values, cultural identities). In this case, the political institutions are known to the analyst and the social variables considered as ready-made categories.[31] The sociology of social problems and some branches of the sociology of social movements tend to adopt similar approaches. The unit of analysis is, in this latter case, the individual behavior of the actor (or the social group), which is supposed to be linked to a certain interest (making his group grow, and "frame" the problem for that end).[32] Here, the analysis tends to take for granted the problem itself, of which only the modalities of its "framing" are modified by the actors involved, as it evolves from an individual concern to a collective issue, possibly through "means of amplification."[33] Taken to its logical conclusion, such an approach is a "social constructionism" that would use social categories ("values" or "interests") as explanatory factors for the evolution of problems.[34] Separating between the "problem" and social categories that are taken for granted ("values," "culture," ...) as well as other types of separation (e.g., between "problem stream" and "solution stream"[35]; among "principles of selection," "culture and politics," and "organizational characteristics")[36] would lead us to identify what is stable enough and can serve as an explanatory category to account for the particular format of the problem. Thus, the nature of nanotechnology's problems could be explained by the power of political institutions in search of new labels for attracting public funding, or by the influence of business actors eager to extend their markets. Political institutions and private companies were indeed important actors in the making of nanotechnology as a science policy priority, as already shown. Yet what also emerged from the example of Nano2Life is that nanotechnology directly raises issues related to the functioning of public institutions, the definition of industrial strategies, the choice of appropriate channels for scientific and political representations. As such, it questions the functioning of institutions, and the modalities of interventions of actors such as governments or companies. "Interests," "values," or "cultures" are certainly at play in the making of nanotechnology. But they are part and parcel of nanotechnology objects, futures, concerns, and publics. They are inscribed in instruments (such as roadmaps), are discussed in public offices or on the streets of Grenoble, and are put to the test with nanotechnology. As such, using them as ready-made causal factors would prevent us from pursuing an exploration that could establish both how nanotechnology is assembled and how democratic issues are raised. Two analytical objectives should be kept in mind to do so. First, one cannot hypothesize that every component of nanotechnology is equally flexible. Thus, the analysis should help display the asymmetries among what is stabilized (e.g., institutions reasserting their strength thanks to nanotechnology) and what is not, and among various problematizations of nanotechnology and democracy (asking questions such as "how do critical social movements succeed, or not, in countering dominant problematizations?"). Second, nanotechnology should be considered as a focal point through which much broader phenomena are concentrated and made visible. Not separating "nanotechnology" from "institutional frames" (or "national cultures," or "economic interests") in order to describe the trajectory of the former according to the characteristics of the latter does not mean that they have equal nature and strength, but that analyzing the constitution of the former is a way of understanding the stabilization and destabilization of the latter. As the following sections will show, the works of Michel Foucault and recent developments in STS are particularly useful to undertaking this program.

Foucauldian Problematizations

The second volume of Foucault's *History of Sexuality*, *The Use of Pleasure*, focuses on the "moral problematization of sexuality in Ancient Greece" (Foucault 1984). As it emerges through this book, problematization is the range of ways to tackle a problem. It comprises the mechanisms through which a question becomes a problem, enters "the domain of true and false," is discussed and dealt with through discursive and/or institutional response. In *The Use of Pleasure*, Foucault seeks to understand how sexual behaviors enter moral or ethical domains and how particular identities and modes of treatment are attributed to problems. The initiative is part of a reflection on the "history of thought," which opposes, for Foucault, that of "behaviors," as well as that of "representations." Writing the history of moral codes or the history of "real" behaviors means basing the analysis on a dualist approach separating the rules and the ways of applying them. Similarly, a history of representation would separate an underlying content from its "representations," and question the adjustments between the two. On the contrary, the analysis of problematization brings the two sides together: while considering the formulation of questions and the expression of their answers, in discourses, texts, and power practices, this analysis seeks to avoid separations between "reality" and "representation," or between "institutions" and "problems." "Problematization does not point to the representation of a pre-existing object, neither does it mean the creation of a previously non-existing object by discourse" (Foucault [1984] 2001a, 1489; my translation). Rather, it describes the ontological making of reality, including, in Foucault's work, the human subject himself.[37]

Within Foucault's project of the history of thought, the objective of the analysis of problematization is to make explicit the general shape rendering the expression of a certain range of solutions possible, and thereby constituting "objects for thoughts" (Foucault [1984] 2001a, 1489; my translation). Thus, the analyst of problematizations seeks to describe the conditions of possibility of certain qualifications of questions, the way through which they can be transformed into problems for which solutions could be proposed. The whole process is a collective production; it constitutes the "specific work of thought," which cannot be separated from the practices and technologies through which it is enacted. Problematization thus defines "the conditions under which possible answers can be provided. It defines the elements that constitute what the various solutions attempt to answer" (Foucault [1984] 2001b, 1417; my translation). But such a formulation should not lead us to think that problematization refers to an underlying structure determining the forms of thought. As Paul Rabinow said, the study of problematization is neither a history of ideas, nor an "analysis of an underlying system of codes that shows a culture's thought and behavior" (Rabinow 2003, 45–46). Rather, again in Rabinow's words, problematization refers to the processes through which a situation is seen "not as a given, but as a question" (18).

What I take from Foucault's work, more than a ready-made concept that could be "applied" to yet another situation, is an attention to the operations of definition of problems and solutions, of ways of thinking and organizing the world, which does not separate a "real" object from its "implications" or "attitudes" about it. As nanotechnology is a loose connection among objects, publics, concerns, and futures, the analytical approach cannot distinguish "nanotechnology" from its "democratic dimensions." This does not mean that there is no distinction whatsoever between "nanotechnology" and its "representations," "implications," or "concerns." But analyzing the problematization of nanotechnology implies that these distinctions are outcomes of processes that need to be empirically accounted for, and which ultimately contribute to problematize nanotechnology in contingent ways.

The focus on problematization allows me not to consider an a priori dichotomy between "nanotechnology" and "problems of nanotechnology." There is no interest here for the separation between the "reality" of the problem and its "framing" or "amplification." Taking inspiration from Foucauldian problematization, I use the concept in order not to posit any distinction among the operations meant to construct nanotechnology as a set of material objects, expectations about the future, concerns to be dealt with, and publics to engage, while exploring the stabilization of its problems. Problematization allows me not to differentiate between "modes of governance" and nanotechnology itself, study the varieties of the coproductionist idiom, and translate them into a focus on the construction of public problems. The problematizations I will look at are "public," in that they are made visible for the analyst himself[38] and are explicit for the making of collective ordering and individual agencies. As such, "problematization" directs the attention to the reception side of the making of nanotechnology, by pointing to the work needed to construct its publics, whether collective or individual. In turn, the public dimension of problematization prompts one to ask many questions. Where are the problematizations visible? How to account for their extension? How to describe the production of social and technical categories through problematization processes?

Processes of Problematization

Foucault considers problematization as processes, stabilized but nevertheless permanently reenacted in order to maintain the definition of problems and devices expected to deal with them. Problematization, for him, is never a given state of affairs, but refers to something that is constantly a problem, for which there is a constant need for solutions and acceptable behaviors (Foucault 1984, 32). Thus, "institutions," "precepts," and "theoretical references" are necessary for problems to be "permanently reformulated." It is then possible, as Foucault describes in *The Use of Pleasure*, to make sexual behavior a problem of measures, of individual ethics, of interrogations linked to everyday practices such as food habits. Foucault demonstrates that the problematization of sexual behaviors in ancient Greece manifests itself by a constant work of writing and reflection, and considers that the production of the technologies of regulation of sexuality occurs in the same move as that of sexuality itself.[39] Consequently, the nature of problematization is an open question for research: one cannot posit that the problematization of sexual activity in Ancient Greece covers the same ground as what today constitutes sex and desire (thus, Foucault insists on the strong link between food practices and sexual behaviors in the problematization of desire as a matter of individual ethics). In this perspective, problematization is a process that shapes a question as a problem, qualifying it, links it with other domains of human activities, and defines a range of potential solutions to undertake.

Thus, the stability of problematization for Foucault is the permanently challenged outcome of a never-ending stabilization process, through which the definitions of problems, the set of potential solutions, and the repertoires of acceptable solutions are maintained. Hence, Foucault's concept of problematization pays attention to the processes that stabilize social order, that provide answers to constantly asked questions. It forces us to consider the institutional, material, and cognitive infrastructures that ensure that problems are stabilized. This directly relates, in the case that interests me here, to Science and Technology Studies in so far as analyzing the problematization of nanotechnology requires the description of technical objects, new technical programs in science policy offices, emerging ethical or risk issues, or concerned publics. I draw on two streams of work in STS in order to ground the further explorations of nanotechnology, namely the Actor-Network Theory (ANT) school and a body of works influenced mostly by Sheila Jasanoff that has insisted on the crucial role of science and technology for the making of state and state-like democratic institutions.

ANT Problematizations and the Emergence of New Issues

Whereas Foucault was concerned with the stabilization of problematizations, STS scholars of the ANT school are more interested in the emergence of hybrid objects (Latour 1991) and matters of concern (Latour 2004), which requires new technical and social arrangements to be dealt with. Following John Dewey, they have recently focused on "issues" to describe the stimulus for the constitution of concerned groups, and new forms of social uptake of public questions (Marres 2007). In this perspective, the issue originates from an entity that acts as an obstacle, as it cannot be dealt with by existing institutions. Thus, Callon proposes to "talk of an issue when the available codes, irrespective of what they are, fail to answer the questions raised by this issue" (Callon 2009, 542). The issue then causes the production of new "concerned groups" (Callon, Lascoumes, and Barthe 2009). To the proliferation of problems is added the proliferation of concerned groups, created for, against, and/or with emerging issues. Problematization appears as the joint result of the mobilization of actors and the evolution of issues, as much as it shapes both of them in turn (Callon 1980, 1986).

In this perspective, problematization describes the continuous work needed to transform new issues into public problems, and their successive evolutions. Faced with a new issue, not taken care of by existing institutions, experiments are introduced in order to make it a public problem for which a range of solutions can be defined. These experiments can be participatory instruments, market devices, price determination tools, or insurance mechanisms. As scientific experiments, they require a material apparatus. As loci where issues are qualified, they are sites of problematization. The analyst's task may then lie in the description of the modalities of the experiments that qualify the problem to be solved. One can then connect this version of problematization with the study of the various components of nanotechnology, as described in the previous section. By directing the attention to the recompositions, the enrollment and translation work needed to interest new actors, the study of problematization, as conceived by the ANT school of STS, suggests analyzing in details the processes through which actors manage to make "nanotechnology" a collective problem—in the case of Nano2Life, a problem of "reduction of fragmentation," of European research policy, of scientific disciplines, of nanotechnology and its concerns and publics.

The ANT perspective helps describe problematization as a gradual process around new issues. It is closely related to an emphasis put on "innovation" as a process through which new entities come to the world and transform it. The Foucauldian perspective, by contrast, focuses on well-established problematizations and the mechanisms that stabilize them. While the two approaches share an anti-dualist stance in their refusal of any dichotomy between "problems" and their "representations," they seem to be opposed according to where they stand relative to their focus on "existing" situations or "new" problems. Nanotechnology, however, forces us to challenge this opposition. It is obvious that nanotechnology mobilizes scientific instruments, technological practices, researchers, and industrialists, in the making of "new" objects (in the case of Nano2Life, nanovectors, nano implants, or nanoparticles used for imagery). In the meantime, existing practices or objects are redefined as "nano," and science policy programs redistribute resources and identify desirable objectives for scientific research. But whether or not nanotechnology is new is a question for the actors involved, whether they seek funding for research projects or programs, or contest the scientific value of the field to criticize the economic interest of market development. Therefore, the analysis cannot take for granted the discourse of the "emerging technologies" that would face "existing institutions" possibly "lagging behind" technological development. This reading would isolate nanotechnology from the set of actors and organizations that actively produce it. For the analysis of the joint production of nanotechnology and democratic order, it is more productive to consider the "emergence" of nanotechnology as the gradual stabilization of a heterogeneous entity relying both on reproduction and recomposition processes. "Problematization," as I use it, comprises a particular definition of the "novelty" of nanotechnology. It is a way to not take for granted the separation between "new objects" and "old institutions" in order to account for different grades in the stabilization of nanotechnology.[40] To restate an expression used by French sociologist Luc Boltanski, it allows the analysis to "escape from the illusion of intemporality as well as from the fascination of the 'new'" (Boltanski 1979, 75; my translation).[41]

Agencements and Problematization

The sites where nanotechnology is crafted are diverse. Consider the case of Nano2Life. The project is composed of dozens of laboratories scattered all across Europe. It is connected to the offices of the European Commission in Brussels and to the European Parliament, where the future of nanotechnology is discussed, and inscribed in science policy instruments such as roadmaps and funding plans. It produces objects and experimental products (nanovectors for drug delivery, carbon nanotubes for brain implants), the regulatory existence of which is uncertain. CEA, its coordinating research institution in Grenoble, is the most important partner in local development projects that attempt to make nanotechnology a key engine of economic growth. It also actively participates in museum exhibits meant to present nanotechnology and its concerns to the local public, while being a prominent figure in the numerous public debates organized about nanotechnology in the area.

In these various places, instruments articulate the components of nanotechnology with each other. Science policy mechanisms define local, national, and European research priorities and thereby connect laboratory practices, visions of future developments, definitions of legitimate concerns, and modes of representation of publics. Participatory devices are expected to make nanotechnology's publics speak and have a say in the public treatment of the risks of nanotechnology objects and their potential ethical issues. Meanwhile, museum exhibits present nanotechnology objects, future developments, and potential concerns to various publics, who are then invited to voice their own opinion. Thus, nanotechnology is problematized through heterogeneous instruments (such as roadmaps, participatory mechanisms, or museum exhibits) that define problems in a public manner. These instruments can be described as "agencement," a notion introduced in recent STS works to describe sociotechnical configurations that distribute agency. Agencements, in Callon's work, are made of material and cognitive elements that shape individual and collective action (Callon 2008). Initially meant to account for the construction of markets, the term helps describe the processes that qualify goods, perform actors' rationalities, and organize collective order (including, but not limited to, the modalities of economic exchange) (cf. Callon 2004 about Barry 2001), in which case they comprise networks of standardization, pricing formulas, audit procedures, and material platforms of exchange. The agencements I am interested in are sociotechnical configurations that problematize nanotechnology, and thereby define and articulate its objects, futures, concerns, and publics. In order to identify problematizations of nanotechnology, I focus on the ways in which agencements make problems explicit and enact ways of dealing with them. They act at both ontological and normative levels, in that they shape objects and actors' rationalities, and define what should be done in the future developments of nanotechnology and for the sake of the democratic organization. The agencements I focus on are made of material, cognitive, and human elements and sustained by more or less formalized expert knowledge. They can be participatory procedures, museum exhibits and accompanying public opinion measures, and processes for the examination of nanotechnology's ethical issues, risk management methods, and forms of social mobilization.

The description of the problematization of nanotechnology through agencements offers a practical path for the analysis of the joint problematization of nanotechnology and democracy. Agencements might be constructed specifically for nanotechnology, or be based on the replication of existing instruments. Consider, for instance, well-established participatory devices (like the consensus conference) or risk assessment methodologies: they are mastered by experts, circulate from one place to another, and are applied to nanotechnology after having been mobilized on other technical questions. They are tools meant to be external to nanotechnology, but nonetheless participating in its problematization. Yet what interests me is precisely the work needed to distinguish them from nanotechnology, or to tailor them to the specificities of nanotechnology, and make them "technologies of democracy," separated from nanotechnology itself (Laurent 2011; see chapter 3).

The analysis of agencements (particularly as they take the form of technologies of democracy) will contribute to the study of "public participation in science and technology," especially those focusing on the production of specific devices, the political organization that they imply, and the way they manage to be sustained. Alan Irwin argues that participatory mechanisms are not ready-made instruments that scholars can evaluate according to their democratic quality, but sites where the public relationship between science and society is enacted through the active, albeit, in some cases, controversial, making of citizens able to talk within particular devices (Irwin 2006).[42] I will follow a similar approach for the study of technologies of democracy, and agencements more generally. Thus, I will consider participatory and deliberative devices as instruments that problematize.

However, I do not limit the analysis to the study of "participatory procedures." This would require an a priori identification of the scope of "public participation" and would prevent from drawing links among devices that nonetheless produce nanotechnology, its problems and publics, albeit in no "participatory" formats.[43] The focus on agencement allows me to adopt a much wider perspective on the sites where nanotechnology is problematized, and, by the same token, democratic order constituted. It questions the very notion of "place," since the description of agencement must account for spatial arrangements.[44] They can be, as in the case of dialogue or participatory mechanisms, public forums where people and ideas compete against one another, but they are not limited to them. Sites where nanomaterials are standardized, where the responsibility of industrialists and scientists is defined, or where nanotechnology is displayed for its various publics rely on agencements that are not "participatory" by nature, but nonetheless contribute, as the following chapters will make clear, in democratic ordering.

Sites and Spaces of Problematization
The description of agencement is localized in the sites where nanotechology is problematized. Yet the analysis has no reason to be limited to a microlevel, for two reasons. First, describing the problematizations of nanotechnology is observing the actors articulating nanotechnology objects, futures, concerns and publics, in sites such as science museums, standardization organizations, or science policy offices, which act as "centers of calculation," that is, as places where connections with many other places are performed. Second, one can also follow trajectories across sites where nanotechnology is problematized. For instance, Nano2Life is connected to the making of European science policy through the circulation of scientists,

administrators, and European officials. In the meantime, concerns and expectations about the future circulate between Europe and the United States, while objects are discussed in regulatory bodies at national and European levels, and international standardization institutions. One can follow objects, as they are produced in scientific laboratories or R&D units, bought by other companies, subjected to regulatory concerns and standardization attempts. From Grenoble to Brussels, and from Washington to Paris, one can also follow the circulations of scientists, officials, and activists.

A classical Actor–Network Theory argument states that the production of the macro is not different in substance from that of the micro, as it is about enrolling actors and stabilizing heterogeneous networks (Callon and Latour 1981). The analysis of agencements indeed gathers both the microprocesses and the construction of macroscopic order. But in order to illuminate the processes of democratic ordering that are entangled within the articulation of nanotechnology's components, one needs to account for differences across types of links among objects, futures, concerns, and publics, variations across geographical and political boundaries, and differences in the connections with political institutions. Consider the case of the Nano2Life project. By many respects, Nano2Life hints at a European way of problematizing nanotechnology, and, thereby, participating in the making of a European Research Area characterized by coordination processes that stem across member states, control mechanisms that do not adopt the form of the legal constraint, and channels of democratic legitimacy that arise from the ability of European initiatives to take "ethical issues" into account and "dialogue" with stakeholders. These ways of problematizing nanotechnology and democracy make little sense if one does not include the European research policy as structured by the Lisbon strategy in the description: they fall into a particular imagination of what constitutes desirable collective futures and acceptable public concerns, based on the association between innovation and (nonelectoral) public participation. The agencements that originate from Nano2Life have value within these European constructs, which they contribute to shape. They participate in the production of a European Research Area characterized by an emphasis put on "competitiveness" and "precaution"—two guiding principles that imply the exploration of issues related to the European identity, to the political organization of the Union, and to its economic strategy (Dratwa 2012). By contrast, they are contested by the initiatives of the Grenoble anti-nanotechnology activists, for whom (as I will discuss in chapter 6) nanotechnology is problematized as an issue for radical critique emanating from an anonymous "simple citizen."

Studying the problematizations of nanotechnology is asking questions related to the production of spaces characterized by common problematizations. These spaces might be outcomes of standardization operations, science policy, or regulatory initiatives. Their examination requires the investigation of connections between the production of small-scale objects in laboratories and private companies, the making of science policy programs endowed with hundreds of millions of dollars and euros, and the construction of shared imaginaries for citizens who will participate in, engage in, voice their concerns about or celebrate the development of a new scientific domain and the introduction of new consumer goods on markets. The study of spaces of problematization needs to focus on the articulation with national and international institutions, in order to illuminate the ways in which they are restabilized or displaced as nanotechnology is problematized. The perspective advocated by Sheila Jasanoff as she analyzes the joint making of the ontological (how are human and nonhuman beings constituted?) and the normative (how are desirable futures defined?) in various institutional settings is particularly useful to ground the approach we need to develop in order to account for the variety of democratic ordering as nanotechnology components are associated with one another. Jasanoff proposes to examine the processes that ensure both democratic legitimacy and scientific objectivity, that make public demonstrations of technical and political proofs accepted at large. This approach compels us to connect these processes with the functioning of political institutions, and, more generally, with state-making operations (Jasanoff 2005, 2012). Translated into our concern for the problematization of nanotechnology, this means examining the mechanisms that make particular problematizations stabilize within particular institutional constructs. In doing so, the point is *not* to study how new problems encounter old institutions, but how agencements gradually stabilize problematizations of nanotechnology, and, thereby, reinforce or displace national or international institutions.

Engaging in the Problematization of Nanotechnology

As social scientists are called on to participate in the definition and conduct of nanotechnology development programs (Macnaghten, Kearnes, and Wynne 2005), they play a central role in the agencements that problematize nanotechnology. Nano2Life's ethics board, for that matter, is just one site of scholarly intervention in nanotechnology projects among many. Numerous research projects are funded by the European Commission or the NNI, while social scientists also participate in the organization of public dialogues, forums, and other participatory instruments (some examples will be provided in the following chapters).

Faced with such an intervention of social science, the analyst can observe the work of social scientists at a distance, and describe the variety of their positions and the modalities of their participation in the problematization of nanotechnology. Analyzing the intervention of the social sciences at a distance would allow me not to take a side in the problematization of nano-technology. It could appear as a way of implementing the agnosticism about democracy that I want to follow. But it is not entirely satisfying either, for both very practical and more theoretical reasons. As I was inter-viewing him about Nano2Life, Patrick Boisseau repeatedly told me that what I did "could be useful." Throughout my research, I had multiple con-tacts with actors (in ways that will be described). I was engaged in policy works about nanotechnology, and with civil society organizations mobiliz-ing on nanotechnology. This is not surprising, as proponents of nanotech-nology programs wishing to integrate nanotechnology concerns are eager to engage social scientists in the making of "more democratic" nanotech-nology. But it causes practical difficulties for the conduct of a scholarly inquiry that would hope to maintain its exteriority, as the actors being studied are very much willing to benefit from its outcomes, or even to take part in it.

There are other reasons, more theoretical, for giving up the stability of the external analytical position. If I were to put social scientists at a dis-tance, there would be no reason for someone else to describe what I am doing when trying to analyze the problematization of nanotechnology (or myself, if I was to adopt a reflexive approach). Once the external position is assumed, it automatically drifts into never-ending introspection. For the descriptions I propose and the connection I draw among sites may contrib-ute to the problematization of nanotechnology. Yet the external position supposes that the reconstruction work is somehow already done, that the social scientist can unproblematically put it at a distance—at the price of endless examinations of problematizations as reconstructed by successive social scientists, each being examined at a distance.

Considering these difficulties, a way out is to deflate the exteriority prob-lem, and to get back to the practicalities of empirical work.[45] In particular, one could consider that the distance between the analyst and the entities he wants to describe is not given from the start, but that it is to be stabi-lized, alongside the other components of the problematizations of nano-technology. For that matter, connecting the problem of exteriority with the focus on problematization is easier if one gets back, once again, to Foucault.

In commenting on Nietzsche and reflecting on the exterior position of history, Foucault provides some directions of thought for the analysis of problematizations that reformulate the question of exteriority (Foucault [1971] 2001c). Foucault's reading of Nietzsche on the question of history makes a critique of the external position of traditional historians explicit.[46] Contrary to them, the genealogical approach he advocates "turns history upside down." It uses fine-grain historical material in order to produce a situated knowledge, which originates from the concerns of the researcher himself, and the contemporary problems she is interested in. Hence, genealogy refuses the separation between subject and object. It does not seek to be (like archeology before it) a discourse putting its topic of inquiry at a distance. Thus, genealogy forces us to consider that reconstruction has more than an analytic component: more than a mere description of the world "as it is," it is also a scholarly reconstruction and a transformation of stable realities into research questions.

This is another reason why the notion of "problematization" as Foucault uses it interests me. On the one hand, problematization is, for Foucault, the state of a discussion at a given time, the ways of defining a problem and the range of possible solutions. On the other hand, problematization is also the outcome of scholarly work, which contributes to introduce in public discussions a particular topic that is thereby denaturalized. Thus, speaking about prison and the role of the Information Group on Prisons (*Groupe d'Information sur les Prisons*, or GIP), Foucault explicitly linked activist action and problematization: "GIP has been a 'problematization' enterprise, an effort to render problematic some evidences, rules, institutions and habits that had been sedimenting for decades" (Foucault [1984] 2001d, 1507; my translation).

These two sides of problematization are not opposed: they are two aspects of the same reality. A dialogue between Foucault and Deleuze offers an illustration of this position. Deleuze links Foucault's work on prison with his own engagement in order to show that the relationship between "theory" and "practice," between "academic work" and "political engagement," is not an issue in Foucault's perspective. On Foucault's engagement in the GIP, Deleuze states: "There was no application, no reform project, no inquiry in the traditional sense. There was something completely different: a system of relays within an ensemble, in a multiplicity of bits and pieces both theoretical and practical" (Foucault and Deleuze [1974] 2001, 1175; my translation).

Thus, accounting for problematizations is also problematizing: the study of problematizations is not a representation of problems independent of the work of the researcher but leads him to connect the successive modes of problems' formulation, up to those that belong to the "concern for the self" of the analyst.[47] The question of exteriority can then be rephrased: the analyst is part of the world he or she studies, and the social scientist looking at problematization is inevitably engaged. But this is not something the social scientist should feel sorry about, and expiate through painful reflexive exercises through which she could locate the "influence" of her "personal interests" on the studies she did. Rather, it allows the researcher to enrich the analysis of problematizations by bringing into the description yet another political dimension—that of her own engagement—and by connecting scholarly work with its normative charge. Accordingly, I am much more willing to follow a stream of thought in STS that asks the field to move "beyond epistemology" by questioning the way it can and should intervene in the world (Jasanoff 1996).[48] As I will describe in the following chapters, the analysis of the problematization of nanotechnology is a way to do so. By the same token, it also compels us to ask a number of questions: how does the engagement of the social scientist empirically play out? How does it contribute to the stabilization of problematizations, or destabilizations of others? How to characterize, eventually, the critical strength of the analysis?

These questions will be explored through the empirical examinations conducted in the following chapters. At this stage, I consider that not only will the types of sites of problematization be diverse, but so will be the formats of engagements. Throughout the following chapters, I will describe the processes through which the reconstruction of nanotechnology occurs, not from the outside, but within the conduct of my description work.

Locating Sites of Problematization

As a heterogeneous entity in the making, nanotechnology is a stake for the organization of the democratic life. It engages decisions about the definition of objects, the planning of future developments, the identification of legitimate concerns, and the representation of publics. By focusing on the problematization of nanotechnology, the objective of the following chapters is to study the coproduction of nanotechnology and democratic order. I propose to describe the agencements that constitute nanotechnology and define it as a collective problem in order to account for the problematizations of nanotechnology. Considering the variety of sites where nanotechnology is crafted and the diversity of problems that are associated with it, it is then important to identify the places where such agencements can be described, and, eventually, common problematizations of nanotechnology and consistent democratic spaces can be identified. The following chapters will introduce some of these sites, in order to reconstruct problematizations of nanotechnology. I choose to group them according to some of the major social operations that make the core of democratic life. Chapters 2 and 3 are devoted to sites where nanotechnology and its publics are represented, in science museums or through technologies of democracy. Starting from issues of representation is a deliberate choice. It will, hopefully, make nanotechnology clearer for the reader without defining it from the start, while also starting to identify differences in French, American, and European problematizations of nanotechnology. Chapters 4 and 5 focus on the government of nanotechnology objects and futures. I examine the controversies about the definition of the "nano-ness" of chemicals in chapter 4, and instruments meant to make nanotechnology development "responsible" in chapter 5. In the last two chapters, I turn to the mobilization within or against nanotechnology, by analyzing two French civil society organizations (chapter 6), and reflecting on the perspectives on the democratization of nanotechnology (chapter 7), both as reconstructed through the American, French, European, and international problematizations of nanotechnology encountered in the previous chapters, and as it appears as a critical approach to the study of democracy. Nanotechnology will appear as a perfect case for the elaboration of a theoretical perspective on democracy attentive to problematization processes, and eventually offering original perspectives for the critical study of constitutional ordering.

Representing

2 Representing Nanotechnology and Its Publics in the Science Museum

Science Museums for Nanotechnology

As nanotechnology became a major science policy topic and a source of public concern, science museums rapidly appeared as sites for the public display of the field. However, there is more at stake in the science museum than the passive representation of nanotechnology. This chapter argues that science museums are sites where nanotechnology is problematized. It describes how European and American science communication experts, museum staffs, scientists, and policymakers debate about and experiment with ways of bringing together the various dimensions of nanotechnology in public representations. In doing so, they question both the nature of nanotechnology and the role of the museum in democratic societies.

This dual focus is not surprising considering the many connections between the science museum and the other sites where nanotechnology is problematized. Science museums are funded by public programs or sponsored by private companies in developing nanotechnology exhibits. They actively intervene in research programs meant to explore the social "implications" or "aspects" of nanotechnology, and are called for to voice the opinions of various publics about nanotechnology. Therefore, these museums are sites where nanotechnology is problematized as a public issue worthy of public engagement, and where, simultaneously, the democratic appraisal of scientific development is put into question.

Science museums—and museums more generally—have always had connections with political institutions. Numerous works have explored the ways in which museums have become places where state power is displayed (Bennett 1995), and where visitors are turned into knowledgeable citizens (Duncan 1995; Macdonald 1996). As they consider the many connections between the public display of science in the museum and the making of the political subject, these works invite us to consider science exhibits and their

accompanying communication tools as agencements. They are instrumented sociotechnical devices that distribute agency to individual visitors
and to the museum as a public institution. The agencement perspective
suggests examining science exhibits not as more or less exact representations of a stable reality, but as heterogeneous devices granting particular
roles for visitors, in the museum space and beyond, and problematizing the
objects they display. This approach is even more fruitful when one considers the recent evolutions of science museums, which directly impact the
forms of public engagement they organize. A first evolution is that "interactivity" has become a central concern of science museums. Studying science museums in France, the United Kingdom, and the United States,
Andrew Barry has described interactivity as a "diagram," making the science museum a crucial site for the formation of a political subject expected
not to be disciplined but authorized (Barry 2001; Callon 2004). The examples that I will describe all make interactivity a central component of the
nanotechnology exhibit. They do so in different ways and for different purposes. In some cases, interactivity is a means for "informal science education." In others, it is a vehicle for the visitor to be directly part of the
representation of nanotechnology, either by contributing to the exhibit, or
to the making of nanotechnology public programs themselves. Interactivity, in some of the cases described in this chapter, acts as a vehicle for a new
channel of public-opinion measure that is expected to have impact on decision making in public bodies.

A second evolution of the science museum relates to a shift from the
representation of science as a black-boxed product toward the representation of science in the making. The former "public understanding of science" objective would now be shifting toward "public understanding of
research" (Durant 2004; Lewenstein and Bonney 2004). Just as the science
exhibit used to display pictures of already-made science, so it now displays
pictures of science in the making. The "Open Laboratory" at the Munich
Wissenschaftmuseum, where visitors can look behind glass walls at researchers working in a nanotechnology laboratory (Meyer 2009) is an illustration
of this trend, which leads science museums to engage in more complex
representations of science, including the display of science as a matter
of controversies opening up social and ethical issues (Yaneva et al. 2009;
Xperiment! 2007).

Both interactivity and the representation of science in the making are
expected by their proponents to contribute to the democratization of science in science museums, no longer under the guise of public instruction
but through the active participation of the visitor, who would discover

science in the making rather than science as a repository of given facts and who would possibly be offered opportunities to voice his or her opinion for public bodies to hear. The examples I will discuss in this chapter are all situated within these broader trends. But they also illustrate variations in the use of interactivity, and in their ways of representing nanotechnology as an entity in the making. These variations make comparison an interesting task. They point to differences in articulation among the initiatives undertaken in science museums, the construction of modes of intervention for the visitor, and wider choices about nanotechnology policy. Eventually, these variations relate to the ways in which science museums are expected to act as actors of democratic life—not only in their extending the scope of who has information about science, but also as they force rethinking about the exercise of citizenship and the legitimate channels for producing collective will. Thus, science museums will appear throughout this chapter as crucial sites for understanding ongoing evolutions of contemporary democracies.

Representing Nanotechnology, Turning Science Museum Visitors into Debating Citizens

A French Science Center in the Midst of Nanotechnology Developments

The Grenoble *Centre de Culture Scientifique, Technique et Industrielle* (Center for Scientific, Technical, and Industrial Culture, CCSTI) is a relatively small science center. About twenty people work full time for the CCSTI, and Laurent Chicoineau, its director, is one of the youngest members of the association of the French science centers' heads.[1] Trained in communication science and in regular contact with natural and social scientists, he outlined during an interview a vision of the science center based on interaction and participation: "So many museums are repositories of objects for the visitors to admire. It is not how I imagine the mission of the science center. For me, the science center is a place where people think about, interact with, participate in scientific research."[2]

For Chicoineau, the science center had a "democratic role to play," which could not be limited to a model based on public education. He saw the redefinition of the role of the science museum in conjunction with the intervention of his science center in the public display of nanotechnology. Since the early 2000s the Grenoble CCSTI has proposed several nanotechnology projects to institutional funders (above all, the regional council) and private sponsors. In 2004, it launched a new exhibit devoted to nanotechnology, which then circulated in the Bordeaux CCSTI and at the Paris *Cité des Sciences et de l'Industrie*.

These initiatives were tightly connected to the policy and industry scenes at local and national levels. They depended on external funding, which were provided by public research bodies and private companies. They included researchers and policy-makers in their design and conduct. In Grenoble, the most visible of these connections was with the Commissariat à l'Energie Atomique (CEA), a national public research institution that had diversified its activities from nuclear energy to the whole range of emerging technologies, and had become a central actor in the French nanotechnology activities, particularly through its Grenoble-based laboratories. Private companies were also included, but their participation was ambiguous. Some of them contributed financially to the exhibit and participated in its design without appearing as sponsors. Others used the opportunity to display their activities in the field of nanotechnology. For instance, a chemical company was a partner of the Bordeaux stop of the nanotechnology exhibit. This company added several panels to the original exhibit, in which it explained why its production of carbon nanotubes was indeed applied nanotechnology, and what its choices were in order to ensure that the production met safety criteria.

The participation in the nanotechnology exhibit was both a financial requirement for the Grenoble science center and a strategic decision for the sponsors. For nanotechnology was a hot topic in Grenoble. Large-scale research nanotechnology projects had been led by the CEA since the end of the 1990s, and had been met by highly visible contestation. Hence, the exhibit was explicitly conceived as an answer to the local anti-nanotechnology activism among other communication initiatives. As it circulated in Bordeaux and Paris, the exhibit was conceived as a basis for public discussions, which took the form of discussion groups involving visitors in Bordeaux, and, in Paris, a two-day public event involving various stakeholders.

New Representations of Science, New Roles for Visitors

A primary concern of the designers of the exhibit was to "connect the representation of the making of nano objects with that of the questions for public debate."[3] The "connection" at stake here was inscribed in some of the devices used in the exhibit, which were meant to display "what nanotechnology does" rather than "what it is." This alternative was regularly mentioned in the preparatory documents. Displaying "what nanotechnology does" related to the epistemological nature of nanotechnology as a scientific discipline based on instrumented practices. The physicist involved in the preparation of the exhibit defined nanotechnology as a matter of intervention on and control of physical matter at the atomic scale, with the help of tools such as the scanning tunneling microscope (STM). The STM pictures individual atoms by displacing them, and thereby renders obsolete the distinction between observation and intervention. It implies coping with physical forces that have different properties than at the macroscale. Therefore, representing nanotechnology was, for the physicists involved, representing how these forces apply. The exhibit's mottos were "seeing through touch"[4] and "seeing and manipulating the invisible."[5] These phrases, which resembled those used by nanotechnology scientists in policy arenas, related both to the nature of nanotechnology as a scientific field and to the set of its potential applications. They referred to interactive devices introduced in the exhibit, and meant to represent nanotechnology by letting the visitor act and experience the action of physical forces at the atomic scale. Some of these devices were quite simple, others more sophisticated. Examples of the former included a boxing glove to be used by visitors to move Lego-like colored objects. They could thus feel what it was like to manipulate matters while being hindered by physical constraints similar to those researchers faced when working at the nanoscale. A more sophisticated tool was a so-called "nanomanipulator" which consisted of a screen on which users could see the moves of a virtual scanning tunneling microscope, and a joystick that visitors could use to move the tip of the microscope and feel the resistance of the atoms thereby displaced—this resistance being quite different from that of macroscale objects because of quantum effects. As the visitor could use the nanomanipulator, he would "notice that nanotechnology was about building, that it was not (...) about picturing reality but really constructing new ones, new applications."[6] The nanomanipulator had been developed by scientific researchers interested in the control of instruments for use at the nanoscale (Marlière et al. 2004). For its designers, the nanomanipulator was supposed to enact a representation of nanotechnology that was not based on the passive representation of nature, but rather was involved in the actual manipulation of objects.

In the nanotechnology exhibit, interactivity was the necessary condition to represent what the physicists considered one of the characteristics of nanotechnology, namely the actual building of matter rather than the representation of a given reality. The representation of nanotechnology that the nanomanipulator enacted meant both displaying and practicing nanotechnology, in ways that mirrored the intervention on which the scientific practices of the field are based.[7] It situated the science center in the midst of the development of nanotechnology. The nanomanipulator was not a mere educational tool for exhibit visitors, but it was also expected to be used by students and experimenters. It was the object of numerous scientific publications (e.g.,Marchi et al. 2005; Marlière et al. 2004) and was circulating in laboratories, as a device expected to train students and scientists in the manipulation of scanning probe microscopes.

Interactivity as performed through the nanomanipulor made it possible to connect the representation of nanotechnology as a technological practice with an interrogation about the potential uses of nanotechnology applications. The visitor was thus expected to initiate his or her reflection about nanotechnology's related concerns. The designing team raised these questions early in the preparation of the exhibit. Eventually, several industrial applications of nanotechnology (e.g., electronic chips, high-performance ceramics, provided by the private companies that were partners of the exhibit) were displayed close to the nanomanipulator. The idea of the exhibit planners was, in the continuity of the nanomanipulator, to use applications as entry points to make visitors think about nanotechnology's future technological developments, but also the future they themselves envisioned.

Practicing and Displaying Public Debate

The representation of nanotechnology within the Grenoble science center articulated the epistemological transformation of scientific practices with the evolution of the expected political role of visitors. Interactivity was meant to transform the nature of representation, the position of the science center, and the role of the visitor. Other components of the exhibit were also participating in this redefinition. For instance, spectacular pictures of nanotechnology were displayed within the exhibit as a way of connecting the representation of science as a laboratory practice and that of science policy as an enterprise producing pictures expected to convince policymakers of the value of the field.[8] The nanomanipulator was accompanied by numerous interactive devices intended to make the visitor reflect on his or her attitude toward technology, such as interactive questionnaires about the use of technology. The questionnaires had been prepared by sociologists, and watched closely by the industrial partners of the exhibit, for whom it raised a marketing interest. It aimed to include the opinions of the visitors to the exhibit, who could then participate in yet another interactive activity. Through these questionnaires, the visitors' opinions became a component of the exhibit: representing nanotechnology was also representing various opinions about technology development. This contributed to one of the main concerns of the exhibit's designers, that the exhibit was supposed to turn visitors into "debating citizens."

Creating these debating citizens is an evolution from a public instruction agencement that the French science museum has long been accustomed to (Bensaude-Vincent 2000; Callon 1998). It is situated within a broader interest in interactivity that national science museums such as the Paris *Cité des Sciences* have been pursuing since the late 1980s. Yet interactivity, in the Grenoble science center, held a specific role. It connected the representation of nanotechnology with that of its debating publics. A technique for the making of debating citizens was a device called *petits papiers* ("little notes") by the organizers. Paper sheets were provided at the end of the exhibit for visitors to leave written notes, which were then displayed as part of the exhibit and examined by sociologists.

The agencement that emerges out of the Grenoble exhibit uses interactivity as a way of integrating the visitor into the representation of nanotechnology. It does not display nanotechnology as a set of given scientific facts, but as an association of objects to be acted upon, imaginaries of future developments, public concerns, and debates. Neither does it attempt to represent "science in the making" or "controversies" as if they could be displayed at a distance.[9] This agencement makes nanotechnology a problem of experimenting with the channels of representation. That the science museum has a "democratic role" to play, as the director of the Grenoble science center believes, does not mean that the visitor could directly contribute to local or national policymaking, but that she is made a debating citizen within the space of the science museum, then better equipped for participating in public discussions.

In later projects conducted by the Grenoble science center, visitors were offered the possibility to design objects meant to be included in public exhibits about nanotechnology, and contests were organized for students to produce films or artifacts about issues related to nanotechnology. I attended the closing session of one of these projects in the spring of 2009. Participants proposed prototypes, films, and scenarios in which they presented what they expected from nanotechnology. Somewhat ironically, the winners were a team of high school students who had displayed in a film an imaginary capsule within which people could live "without nanotechnology." But there is no irony if one situates this intervention within the agencement the Grenoble science center constructed, within which visitors were turned into debating citizens, practicing the debate about nanotechnology at the same time they participated in its display. It was then entirely consistent that the imaginary "nanotech-free" space was rewarded as a contribution to the nanotechnology debate, and a visual proof that living in a world where nanotechnology had been developed was possible even for people who did not want it.

An Experimental Democratic Agencement and Its Critics

There were real people in Grenoble who did not want nanotechnology, however, and these people were critical of the Grenoble exhibit. They were anti-nanotechnology activists—the very people to whom the nanotechnology exhibit was supposed to respond. Activist groups in Grenoble were opposed to nanotechnology research, and were attentive to the activities directed toward publicizing nanotechnology. They had published online texts that directly targeted the Grenoble science center and the communication policy of CEA. They had criticized other dialogue experiments that Grenoble's local elected bodies had attempted to organize, and had set up demonstrations on the construction site of a research center expected to be a major nanotechnology center in Europe.[10] The director of the CCSTI, who expected potential demonstrations, requested "special protection" for the opening ceremony of the nanotechnology exhibit, as he "feared for the safety of the guests."[11] Numerous policemen were present during the official opening event of the exhibit. The activists reacted by pointing to the material display of the connection between the Grenoble science center and the public bodies that, according to them, were supporting nanotechnology development without democratic control.[12]

The Grenoble science center was directly targeted in the fall of 2009.[13] Red paint was projected on its walls, and leaflets were left in front of the main entrance. Signed by a "collective for citizen debate" (*collectif débat citoyen*), they explained that the museum was targeted since it was "a symbol of the acceptabilization [*acceptabilisation*] campaign orchestrated around nanotechnology," meant to "prevent social mobilization" against a technological domain that caused health risks and was developed for economic or military interests. Laurent Chicoineau, the director of the science center, answered on his blog, and clearly situated the locus of the confrontation. For him, being "anonymous," as the collective was, and using "violence" (albeit without much consequence for anyone), was a "curious way to defend democracy."[14] Democracy was, for him, precisely what his science center was doing. Hence the opposition: for the activists, the French science museums could not pretend in any way their activities were intended to ensure a democratic appraisal of nanotechnology. The democratic model that the nanotechnology exhibit was constructing, based on the production of representations by visitors themselves, and on the display and practice of debate within the exhibit or in close connection to it, was not accepted by the activists, who considered that their role, as engaged citizens, was to perform a critique of nanotechnology from an exterior position.

The stability of the agencement making interactivity a condition for the formation of debating citizens was also threatened from the inside. At the Paris *Cité des Sciences*, a "totem"—as it was called—was added to the exhibit. It was a tower with large-scale pictures of nanotechnology applications, illuminated from the inside, facing a pool of water where lotus leaves represented "an example of complete natural molecular assemblages."[15] This two-part addition to the original exhibit led the visitor to be puzzled by the beauty of nature, and even more by the mythical power of science, able to transform nature and make it realize its otherwise silent potentialities. The totem situated the exhibit at an objectifying distance from both nature and science. It became a physical place where the passive beauties of the former and the active marvels of the latter were to be displayed. For the museum staff in Grenoble and Bordeaux, this was at odds with how they had attempted to problematize nanotechnology. The totem made the science center an external place where the visitor was not the debating citizen they had hoped to enact, but rather a passive spectator of science and nature conceived as unproblematic sources of admiration.

Facing external criticisms and internal misunderstandings, the Grenoble science center's attempts at redefining the representation of science and transforming visitors into debating citizens are not grounded on stable institutional infrastructures. They are rather experimental forays into a redefinition of the role of the science center in France, based on the problematization of nanotechnology (and more generally, technological development) as a matter of multiple representations, including that of the public debate. This redefinition makes the problematization of nanotechnology in the French science center quite different from other examples, as those of European projects will show.

From the Representation of Nanotechnology to That of Its European Publics

The Grenoble science center was an active partner in European projects devoted to the Ethical, Legal and Social Aspects (ELSA) of nanotechnology. One of them, called Nanodialogue, was a project that Chicoineau, director of the Grenoble science center, saw as an opportunity to "pursue with European partners the initiatives undertaken in the Grenoble area."[16] Yet he became more and more skeptical as the project evolved, for reasons that will be discussed later. Nanodialogue ended up problematizing nanotechnology in a different way than in Grenoble. As one of the first European ELSA projects and a first step in the development of the European approach toward the communication of nanotechnology, Nanodialogue is a site where the problem of the integration of ELSA into nanotechnology was particularly visible.[17]

The Nanodialogue Project: Interactivity, ELSA, and Public Opinion
Together with the concern for the "ethical issues" that were supposed to be taken care of by dedicated bodies within the European science policy organizations, the stress put on "dialogue" in nanotechnology policy documents makes nanotechnology a case among many others in the European science policy landscape. As a Nanodialogue presentation leaflet of explained: "Engaging citizens in dialogue and discussions about science and technology has been recognized by the European Commission as a fundamental component to create the knowledge economy and the basis of the European Union's Lisbon agenda."[18]

The Lisbon agenda, launched in 2000 by the European Council, had indeed called for the transformation of Europe into a "knowledge-based economy," and of the European public into a "knowledge society." In this approach, "dialogue" among scientists, policymakers, and the European public was an important component. Nanodialogue was situated within these objectives. Early on, the project was meant to be a response to the shortcomings of "traditional modes of government." As opposed to "hierarchical, state-led decision-making processes," Nanodialogue was based on a call for "new forms of governance (...) based on networking among stakeholders, on the integration of interests, and on the involvement of citizens and consumers in the implementation of policies." (ibid, 4). The project was based on the hypothesis that public participation had value, in a context described as that of great public concern for the potential implications of scientific research.

A team of sociologists, led by Simon Joss, participated in Nanodialogue. Joss, an internationally known specialist of public participation,[19] had written on consensus conferences, was participating at that time in another European project called CIPAST to train officials and academics in the practice of public participation in science and technology, and was interested in the "democratic ambition" of the Nanodialogue project. He made that clear to me when I met him for an interview: "At the time I thought 'well this is really innovative.' (...) It's a knowledge transfer project where educationists, museum specialists, social scientists, and technology experts come together and try to explore the development of new types of interaction. (...) I thought 'it's exciting, you can do something. Maybe you can work on, you know, democratizing nanotechnology.'"[20]

When the project started, it was evident for everybody (whether partners within the project or program officers in the European Commission's Directorate-General for Research and Innovation) that "democracy on nanotechnology" was to be constructed, and that it had to be done in conjunction with the examination of the ELSA of nanotechnology.

The "democratic component" was an object of discussion among the project members, who considered that the Nanodialogue exhibit was to "make people realize that they were taking part in a democratic process."[21] As in the Grenoble exhibit, interactivity was explicitly linked with a democratic ambition. But whereas the Grenoble science center connected interactivity with the problematization of nanotechnology as a scientific practice blending representation and intervention in the physical as well as in the social world, the problems raised by Nanodialogue revolved around the nature of nanotechnology's ELSA and the way of integrating its aspects within the exhibit.

Within Nanodialogue, the problems of representing nanotechnology in the science museum were related to the appropriate level of content related to nanotechnology's ELSA. Participants in the project argued over the treatment of the original focus on the "societal implications" of nanotechnology.[22] These concerns eventually led the designers of the exhibit to add panels on the "risks" and "ethical" issues of nanotechnology. Some participants in the project considered that this addition was too superficial. But for others, nanotechnology's ELSA was far too visible in the exhibit. The Italian coordinator thus explained during an interview:[23]

We had contacts with scientists. For instance those we work with here. And many of them thought it was way too much about the risk and ethical issues. (...) Cos', you know, ... all the exhibit would say: "there are biomedical applications," and then "and there are all these ethics questions"; "there are these daily life applications, like energy storage," and then "but technology might have safety risks."... And for many scientists, that was just too much insistence on the "ELSA" part, it was not about nanotechnology at all.

This last quote is revealing. It shows that the discussions about the representation of nanotechnology in the science museum within

Nanodialogue shifted toward discussions about the appropriate ELSA component. Representing nanotechnology in the Europe science museum became representing its ELSA. But while the connection between visitors' opinions and policymaking was never an issue in Grenoble, the focus on ELSA was complemented by devices expected to make the European public speak, and expected to realize the democratic ambition the project had been based on.

From the beginning of Nanodialogue, the production of recommendations meant to be transferred to the European Commission was indeed an objective. These recommendations were eventually produced through focus groups, coordinated by the team of sociologists involved in the project and led by each participating science center. These focus groups were meant to present the "viewpoint of the European citizen on nanotechnology" to the European Commission.[24] The recommendations eventually presented to the EC were mostly general lessons compatible with the Action Plan. They insisted on the necessary "precaution" to adopt in order to develop nanotechnology, and identified more "benefits" than "risks." They were not considered as more than a "snapshot" by the sociologists involved.[25] One could easily identify the ways in which the guidelines of the focus groups distributed to the participating science centers determined the final outcomes.[26] But what matters here is less their unsurprising content than what they say about the problematization of nanotechnology that resulted from Nanodialogue: nanotechnology was both a matter of ELSA and an issue of public opinion.

At the final conference of the project in the European Commission headquarters in Brussels, it became clear that the European public opinion was to be measured in more sophisticated details. Simon Joss argued that it was important "to develop notions of the publics, in plural terms, to recognize that the public comes in different forms and shapes and that therefore developing governance modes needs to recognize there's a plurality of the public."[27] This call was just one manifestation of a more general concern for the connection between the problematization of nanotechnology in the terms of its ELSA, and that of European publics expected to have a say about the development of nanotechnology.

Nanodialogue as an Experiment for the European Nanotechnology Communication Policy

As one of the first European projects in both nanotechnology communication and nanotechnology "societal implications," the Nanodialogue project is of particular interest because it served as a rehearsal of the European strategy in nanotechnology communication. As with other projects devoted to nanotechnology's ELSA that put an emphasis on dialogue,[28] the Nano-dialogue experience circulated widely in the communication of the nano-technology and converging science and technology unit in charge of the European initiatives in nanotechnology at the Directorate-General for Research and Innovation of the European Commission. Nanodialogue was presented repeatedly at international conferences on science communica-tion, and various European initiatives made use of the project. For instance, CIPAST, the European training program in participatory instruments, had participants discuss Nanodialogue under the supervision of Simon Joss. At this point, the project had become the topic of a typical case that could be used as an example presented to would-be organizers of participatory devices.

The conclusions of Nanodialogue were supposed to feed the further con-struction of the EU policy on nanotechnology. Immediately after the final conference of the project, a workshop was held in Brussels that gathered proj-ect participants, European officials, and experts in science communication. The workshop resulted in a working paper on developing a strategy for com-munication outreach in technology (Bonazzi 2007). This working paper was later refined and developed into a document written by Matteo Bonnazi, offi-cer at the DG for Research and Innovation of the European Commission.

This report, entitled *Communicating Nanotechnology* (Bonazzi 2010) out-lined the "communication roadmap" that was to frame the strategy of the European Commission on the communication of nanotechnology. This strategy was based on a "new mood of communication (...) based on dia-logue" and the report stipulated that "instead of the one-way, top-down process of seeking to increase people's understanding of science, a two-way iterating dialogue must be addressed, where those seeking to communicate the wonders of their science, also listen to the perceptions, concerns and expectations of society. (...) This should enable to settle a sound basis for reaching consensus, achieving sustainable governance and social accep-tance for nanotechnologies and nanosciences" (Bonazzi 2007, 10).

The report thus pursued some of the issues that had been central in the Nanodialogue project, namely "dialogue" and the evaluation of "percep-tions, concerns and expectations of society." It considered science commu-nication "as part of the research process itself."

That nanotechnology communication was "part of the research itself" was rendered possible—at least institutionally—by the fact that the man-date to the European Commission defined a "double role for the Nano and Converging Sciences and Technologies Unit" (the expression was

Bonnazzi's[29]) in the Action Plan. The unit was expected to define calls for scientific research projects, and, at the same time, had to work on communication. Crafting communication coming "from the very core of research," as Bonnazzi said to me, implied that the Nanotechnology and Converging Sciences and Technology Unit at the EC's Directorate-General for Research and Innovation was also in charge of "science and society" topics, as the reorganizations of the DG had just made possible.[30]

The roadmap for nanotechnology communication defined the "goal of communicating" as a "gain in EC image," particularly as far as "transparency, credibility and accountability" were concerned (Bonazzi 2010, 71). The hope was that the "consensus-based support to EU policy-making on responsible nanotechnology within society" could be increased (ibid.). In order to do so, the roadmap proposed extremely simple messages to convey:

Nano is: **not** magic;

Nano is: **a new phase of technology** exploiting nanoscale effects;

It deals with new: **beneficial applications and markets**, impacting on **health, safety, privacy, ethics, and the socioeconomic divide**;

It: **must and can** be controlled and driven conscientiously. (Bonazzi 2010, 106; emphasis in the original)

For all their simplicity, these messages also insisted on some of the main focuses of Nanodialogue, namely ELSA and the fact that nanotechnology was a program open to conscious direction. Eventually, the content of the "main message" was not the most problematic point of the roadmap, which considered nanotechnology as either a set of scientific objects and domains that could be described, or a source of potential uncertainties that raised ELSA aspects. Rather, all the work to be done was to identify potential "targeted audiences" (e.g., "youngsters," "media" or "NGOs"), potential communication techniques (primarily "two-way methods" such as "dialogue" and "participatory" devices), and linked the first with the second. Instead of developing the ways in which nanotechnology could be represented in science museums, the bulk of the "communication roadmap" was about distinguishing between types of audiences (e.g., "children," "youngsters," "scientists," "NGOs"). It could then provide synthetic tables of European initiatives in the communication of nanotechnology, which were classified according to their "targeted audiences." From Nanodialogue to the European roadmap, the main concern had shifted from the representation of science to that of its publics.

This shift was described, in the roadmap, as an evolution "from 'public understanding of science' to 'scientific understanding of the public.'" This move implied that the "public" was to be scientifically known, in ways that also allowed "dialogue" and "exchange of information." Dialogue, in this model, is used as a way of getting knowledge about the public, to be attentive to its "expectations and concerns." It is an instrument in which the "main message" to communicate to the public is not questioned. Hence, the "scientific understanding of the public" tailors the activity of representation no longer toward nanotechnology, but to a European society whose interest in nanotechnology needs to grow. For the head of the Nanotechnology and Converging Technologies Unit at the DG, what was to be constructed through the "scientific understanding of the public" was nothing less than "technical democracy": "These tools will allow a technical democracy platform to be put in place: public opinion will be monitored on a continuous basis through Web-based measures that could be picked up by other media. (...) (They) will make the platform one of the most appropriate means to monitor what people really think about nanotechnologies and promote evidence-based dialogue" (ibid., 152).

Here, the "evidence-based dialogue" is not problematic because of the representation of nanotechnology but because of that of "the public." "Continuous monitoring" can thus appear to solve the "problem of representation" (an expression used by an EU official during an interview) that EU officials have regarding the organizations from civil society they are in contact with. One of them noted: "That's an issue here, it's always the same kind of people, over and over again. We do a meeting open to civil society, we request comments. . . . And we can guess in advance who's gonna show up. They're always the same, Friends of the Earth, maybe Greenpeace, ... And what we want is talking to the European public, to the real European public."[31]

Defining the "real European public" of nanotechnology and the infrastructure able to make it speak to the European institutions is an important issue. It problematizes nanotechnology in ways that define who is entitled to speak to the European institutions, and for what results. For that matter, the European civil servant in charge of nanotechnology who voiced this concern for the "real European public"[32] was skeptical about the value of "dialogue," if it was to be held with established stakeholders. What made the public "really European" was, for him, less the fact that participants in dialogue knew and mobilized on nanotechnology, as many NGOs intervening in the debates about the European regulation of nanomaterials do (see chapter 3), than their being "as diverse as the European society is."

Making (Nano)technology Research European

The ongoing process intended to provide continuous feedback of public opinion on nanotechnology has several objectives. The Directorate-General for Research and Innovation hopes to be able to correct the misrepresentations of the public, but also to develop certain areas of nanotechnology rather than others. Talking about a call for project he was crafting, a EU official at the DG recently explained during an interview:

> If we are not able to give the possibility to the public that is participating in the dialogue to really see that what they are dialoging on is put into concrete policy action, there's no need. So if at the end of the story we have a book, it's a failure. So the condition I'm putting in this call is the following one: that the successful projects (...) will provide evidence that there is a link between what is being discussed and what is going into the changing, or re-addressing, or reinforcement of the current EU policy. That means on current funding lines for nanotechnology. I'm putting this as a condition, it's something quite new that engages not only the public but also ourselves, the regulators. (...) So, for sure, the main input of this will be on funding research. So if the public, or those publics, or different member states, say to us "please don't do research on nanofood," we will not spend any single euro on nanofood.[33]

Nanotechnology forced the DG to refine the representation of nanotechnology: as a science policy program defined by the amount of funding it was granted, as a topic of potential public sensitivity, the issue with how nanotechnology is represented became less that of the representation of science than of the correct representation of public opinion. It is in that sense that nanotechnology is an opportunity to construct a "technical democracy."[34] In this technical democracy, the scientific understanding of the public (rather than the negotiation among stakeholders) is expected to contribute to the making of European nanotechnology policies. In this process (and one can trace it back to the early European project on the "societal implications of nanotechnology"), the scientific representation of the public is built on the exact same theoretical basis as public understanding of science: the problem is to ensure the faithful, at-a-distance representation of an object the existence of which is not problematized. Thus, the initial interrogations about the "democratic ambition" of nanotechnology policy that were made explicit during the Nanodialogue project appear to be solved: the "scientific understanding of the public" is expected to connect the European nanotechnology policy with its publics, and the whole process implies shifting from the representation of nanotechnology to that of the European public. The "democratic ambition" thereby translates into the production of new channels of political legitimacy: the representation of nanotechnology and its implications need to be ensured, while the scientific representation of the public is expected to ground the formation of a European polity.

Hence, it is now possible to better understand the idea of integrating nanotechnology communication "at the core of scientific research." This integration implies the problematization of nanotechnology in the terms of the examination of its ELSA, and the representation of the "European public." It is based on a well-specified distribution of roles, where the Directorate-General on Research and Innovation of the European Commission needs to gather information about the nature of nanotechnology's ELSA (possibly through social scientific expertise) and public opinion about potential science policy options. This implies an institutional evolution making it possible for science policy offices to deal with "science and society" issues, and also that European nanotechnology policy has the capability to react once a sign of social concern is perceived, either to commission risk studies, or to redirect funding to certain areas rather than others.

An American Expertise in Informal Science Education

In March 2009, I met Margaret Glass, the coordinator of a network of American science museums involved in nanotechnology activities—the NISE (National Informal Science Education) network. When she learned that I was interested in the connections between science museums and nanotechnology policy, she immediately compared the American museums with their European counterparts:

In Europe, (...) policymakers want to listen to what people say. Science centers have a real grip on nanotechnology governance, you know, and the EU wants them to help them ... you know ... help identify what people's concerns are. We don't have, for instance, Nanodialogue where the EC set that up and asked for recommendations about policy. That's the missing link in the U.S., we have no feedback mechanism to policymakers. I mean we can present (something) to them, but then they'll have to listen. And they're not asking. The difference is that nobody has asked us.[35]

The difference between the NISE network and the European approach to the role of science centers in nanotechnology policy seemed clear for her. Whereas European policymakers were funding science museums to represent nanotechnology for the public, as well as, if not more than to represent public concerns and expectations for policymakers, she felt that the American science centers were isolated from the actual making of American nanotechnology policy. The roles of the American science center and the

problematization of nanotechnology that it enacts are indeed quite different from what we have encountered so far. In American as in European
science museums, interactivity is heralded as a necessity, and the representation of nanotechnology as a scientific field is discussed. But the American
museums, through the NISE networks, problematize nanotechnology neither as an issue of representation of a heterogeneous entity in the making,
nor as a matter of ELSA, but rather as a distinct scientific domain for which
"informal science education" is required.

Representing Nanotechnology through the NISE Network

Reflections on the representation of nanotechnology occurred at an early
stage in the construction of U.S. nanotechnology policy. In September
2004, a workshop organized by the National Nanotechnology Initiative
(NNI) was held in Arlington to explore the "opportunities and challenges of
creating an infrastructure for public engagement in nanoscale science and
engineering." (Chang and Semper 2004). The workshop gathered about fifteen science museum representatives, and NSF high-level staff, including its
director, Mihail Roco. "Public engagement" was indeed considered a "priority" for the federal program, since the "societal issues" make it "critical for
NSF" to "engage public audiences" (ibid., 4). Indeed, the whole workshop
was structured around the various audiences that needed to be taught about
nanoscale science and engineering ("teachers," "K-16 students," "general
public," "workforce," "community and public leaders" and "scientists")
(ibid., 7). The division according to "audiences" is familiar: we already saw
it at play in the case of the European nanotechnology communication
roadmap. Yet the perspective was quite different in the 2004 Arlington
meeting: the workshop mobilized the various concepts of the so-called
"deficit model" that the European actors were keen not to use. The objective was to "reduce irrational fears," foster "nano interest" and "nano liter-
acy," in a context where the American nanotechnology program needed
students, workers, and consumers (ibid.).

   This definition of the problem of public engagement in nanotechnology
was consistent with the objective of a network of museums specialized in
"informal science education." In 2003, four museums of science (the Boston Museum of Science, the Exploratorium in San Francisco, the Science
Museum of Minnesota, and the Oregon Museum of Science and Industry)
gathered within the Network for Informal Science Education (NISE) received
$750,000 of funding for the following fiscal year, with the objective to "pro-
mote public understanding of nanoscale science and engineering concepts,

scientific processes, and applications to society. The purpose of these efforts is to ensure that the public is kept abreast of advances in the field."[36]

The focus on public understanding of nanoscale science and engineering led program officers at NSF to raise issues about how to represent nanotechnology in the science center. They insisted on the work needed to represent "how size can make a difference in the properties of materials," but also to "appreciate the interdisciplinary nature of nanoscale science and engineering" (ibid.).

Other partners then joined the four initial NISE members. In 2009, about twenty museums were involved in the activities of the NISE network, which had received more than $20 million from the National Science Foundation for five years of funding.[37] Contrary to the projects we have encountered so far, the NISE network was not conceived around the collaborative design and staging of exhibits. NISE is above all a coordination tool that allows American science centers to share exhibit modules about nanotechnology developed by some of the partners, and methods and tools for "public engagement in nanotechnology." The network also distributes ready-made layouts of oral intervention, such as an "Introduction to Nanotechnology" speech, with associated PowerPoint presentations. Each of the components of the NISE production is accompanied by standardized evaluation grids, which, once filled out, are used by the network to refine its offers. The most important common event organized under the NISE umbrella is the annual *Nanodays*, during which activities and exhibits are organized throughout the country in science centers. During this week-long event, which in 2009 involved about two hundred science centers across the United States, highlights include displaying nanotechnology applications, organizing children activities such as building a human-sized model of carbon nanotubes, holding public conferences, and distributing stickers that read "I'm made of atoms."

NISE was funded, within the NNI, through the Nanoscale Science and Engineering Education Program.[38] The NNI emphasizes "informal" alongside "formal" educational activities. This is what a brochure published by NISE argued:

One benefit of a more scientifically literate public is increased support for funding of research. A substantial majority of Americans support government spending for scientific research, including basic scientific research. The better our research and its implications for society are understood, the better the general public can make responsible decisions about public funding. (....) Another motivating factor is to encourage the next generation of scientists. We need children to consider and pursue careers in science and engineering.[39]

Hence, informal science education too could transform the visitor to a science center into a potential supporter, a future scientist, a citizen participating in her country's political life, or a consumer of nanotechnology products. This implied developing ways to make sure it could happen.

Representing Nanotechnology for a Responsible Citizen

The first task of the members of the NISE network was to identify the "important messages" to convey to the American public. Crafted with the help of a group of scientific advisors, the "messages" were eventually the following:

Nanoscale effects occur in many places. Some are natural, everyday occurrences; others are the result of cutting-edge research.

Many materials exhibit startling properties at the nanoscale.

Nanotechnology means working at small-size scales, manipulating materials to exhibit new properties.

Nanoscale research is a people story.

No one knows what nanoscale research may discover, or how it may be applied.

How will nano affect you?[40]

One can compare these "messages" with the multiple representations of the Grenoble nanotechnology exhibit, and the stress put on the ethical, social, and legal implications of nanotechnology in the European projects. They did not hint at the diversity of nanotechnology (comprising industrial applications, science policy programs, public concerns, or debates) represented through the multiple channels of representations in the Grenoble exhibit. Nor did they focus on the ELSA the European science museums were so concerned about. Indeed, the "messages" developed and supposed to be transmitted through the NISE network partners were all about "what nanotechnology really was" in order for the visitor "to make up his mind and act as a responsible citizen."[41] The reality of nanotechnology, then, was about the "nanoscale": nanotechnology was only characterized by the atomic scale of observation and action. Therefore, it made no sense in this perspective to inquire into the collective construction of objects and concerns (as in Grenoble) or into the direction of science policy programs (as in Europe). The nanoscale was a domain out there explored by scientists (and this is the reason why it was "a people story") who entered a new world where "no one knew what would be discovered."

Accordingly, what was supposed to be provided for the citizen was reliable scientific information, rather than reflections on the potential impacts on nanotechnology.[42] Consequently, the productions of the NISE network (which are rather those of each separate partner) focused on the correct description of nanotechnology research practice and industrial applications. The collaboration with science laboratories was heralded as a key objective of "informal science education," both for scientists to use expertise about how to communicate to the public, and for museums to make sure the scientific content of their exhibits and activities was consistent.[43] Ready-made exhibition components were proposed to the NISE members, with all the descriptions and instructions provided on the NISE website. They were peer reviewed by external scientific advisors, and evaluated by the partnering museums through the web platform, thereby ensuring that "learning goals" were met. For instance, the NISE website presented an "Introduction to Nanotechnology Exhibition" proposed to instruct visitors that "things at the nanoscale are super small," "super small nanoparticles can have very unexpected properties," and "scientists are figuring out how to create and manipulate materials at the nanoscale through self-assembly."[44] Different media were used (texts, interviews with scientists, animated films) and interactive devices were proposed. For instance, the "Billion Beads" activity proposed: "Visitors inspect tubes that hold quantities of one thousand tiny beads, one million beads, and one billion beads. To the naked eye, the tube containing one thousand beads appears nearly empty. Visitors see that the next tube, partially filled, contains one million beads. Finally, to compare, a four-foot tall container nearly full contains approximately one billion beads."

Hence, the interactivity that the NISE exhibit proposed was quite different from the direct involvement of visitors in the practice and making of Grenoble nanotechnology exhibit. Interactivity was a means to produce an individual citizen knowledgeable enough about nanotechnology, understanding the "basic facts," and who could then act as an enlightened voter or consumer—possibly a supporter of nanotechnology. Hinting at the ethical issues (as in Nanodialogue) or the "nanotechnology debate" (as in the Grenoble nanotechnology exhibit) was never an issue for the NISE partners. The Grenoble exhibit considered various ways to define nanotechnology. Nanodialogue was all about reflecting on nanotechnology's ELSA and considering the domain as a public issue on which the opinion of the European public was to be gathered. The American science centers and their leaders considered that nanotechnology was a science before anything else, and that it was their duty to represent it as such.

The "New Mission" of Science Museums

The idea of "dialogue"—so prominent within their European counterparts—was not foreign to the American museums, however. A NISE publication targeted to scientists stated that the "monologue style of communication" had failed "to win public trust," and that they need to "move from a 'monologue' model of communication, with scientists lecturing the public on what it should know, to a 'dialogue' model, in which scientists meet the public in forums that are evenhanded, giving nonspecialists much more time to air their concerns and share them with the 'experts.'"[45]

Larry Bell, a co-director at the Boston Museum of Science and principal investigator of the NISE network, spoke in 2008 of the "new mission" of the science museum (Bell 2008; see also Reich et al. 2007). For him, the new mission consisted of ensuring that the public of the science museum was engaged in "two-way communications" with experts and scientists. Bell elaborated his idea of this mission accompanied by the development of a mechanism at the Museum of Science called a "forum": a series of presentations by invited speakers in front of a self-selected audience, followed by several rounds of discussions among the participants divided in small groups. In the first series of forums organized in 2006–2007, participants discussed nanomedicine and nanotechnology applications for energy. In 2008, two forums at the Boston Museum of Science aimed to directly contribute to the decisions of the Cambridge City Council. During these forums, participants talked about the potential regulation of nanotechnology research in Cambridge, and the oversight of the risks of nanoparticles. They engaged in discussions about "municipal oversight of consumer products made through nanotechnology," through exchanges on a series of consumer products. They were then invited to vote on predefined options, such as "should citizens/consumers be made more aware of the lack of research on the safety of some nanoparticles in consumer goods?" or "should there be warning signs or labels?"[46] In Cambridge, where active public involvement in local decisions about science and technology has historical precedent,[47] the staff of the Museum of Science considered the forum a way to make public deliberation "relevant" for policymakers. Local city councilors were regularly invited, and the forum conceived as contributing to reflection on the local regulation of nanotechnology research.

Since its early uses in Boston, the forum has circulated across American science centers. As it started to be used in more and more places, its objectives also became less clear, and spurred numerous discussions about their integration within the informal education strategy of the NISE network.

The uncertainty about the role of the forum was visible as NISE members gathered to discuss the organization and standardization of the forum format. I observed one of these meetings, at the Boston Museum of Science in January 2007. This three-day closed meeting was held as the NISE network was already up and running. Forums had been organized in all the museums that were represented at the meeting (Boston, Minnesota, Oregon, and Raleigh, North Carolina). However, there had not been coordinated actions at that time. The difficulties the NISE members encountered with the forum format became clear. Meeting participants wondered about the connections between the forums they organized or wanted to organize, and policymaking. Some of them questioned the ethical basis of using visitors' contributions to provide policymakers with information on public opinion. For others, what mattered was to make participants in forums influence nanotechnology policies. Still others thought that the forum could not, in its actual form, provide any recommendation, but that the transcript of forum discussions could be handed over to social scientists for them to make sense of the exchanges.

These discussions were all about the uncertain introduction of dialogue as an objective of the American science museum, and the ambiguity about its expected purpose. But in the official documentation of the NISE network, nothing remains of this ambiguity. The forum is described as a ready-made device, with explicit organizational methodology, from examples of discussion topics to practical tips about the food and drink to provide, and sophisticated evaluation grids. As a producer and distributor of expertise about informal science education, the NISE network developed the tools and instruments necessary to standardize the forum into a device aimed to contribute to its objective of informal science education. The standardized forum format is based on a representation of nanotechnology that would at least comprise "basics," explaining, for instance, that "nanotechnology has to do with very small things, smaller than you can see with an ordinary microscope," and that "materials can have different characteristics at the nanoscale."[48] As for the objectives of the forum, they are presented as such in the methodological booklet distributed to the NISE members:

Forum goal
To provide experiences where adults and teenagers from a broad range of backgrounds can engage in discussion, dialogue, and deliberation by:
- enhancing the participants' understanding of nanoscale science, technology and engineering and its potential impact on the participants' lives, society and the environment;

- strengthening the public's and scientists' acceptance of, and familiarity with, diverse points of view related to nanoscale science, technology and engineering;
- engaging participants in discussions and dialogues where they consider the positive and negative impacts of existing or potential nanotechnologies;
- increasing the participants' confidence in participating in public discourse about nanotechnologies and/or the value they find in engaging in such activities;
- attracting and engaging adult audiences in in-depth learning experiences;
- increasing informal science educators' knowledge, skills, and interest in developing and conducting programs that engage the public in discussion, dialogue, and deliberation about societal and environmental issues raised by nanotechnology and other new and emerging technologies. (ibid., 7)

As defined in the NISE document standardizing the methodology, the forum is meant to ensure the public understanding of nanotechnology ("learning experience"), which can be used by the network members to convey the "main messages" defined at the onset of NISE (e.g., "how will nanotechnology affect me?"). Participants can then be good citizens, open to true and balanced information; the "positive and negative impacts," the "diverse points of view" are to be considered alongside scientific information, but are not for the participants to decide upon. Accordingly, the evaluation of the NISE forums is based on the measure of the knowledge the participants have acquired. Evaluation reports of the NISE forums provide sophisticated statistical measures of the "impacts" on the "understanding" of nanotechnology.[49]

Hence, the many discussions about what exactly the "impact on policy" of the forum meant did not result in a construction of a European-like, scientific understanding of the public. Nor did it provide ways for the American museums of science to envision other roles for the participant than that of an individual citizen, consumer, and voter-to-be through the "magic of dialogue."[50] The difficulties the participating museums had to face were dealt with through "deliberation" used as an educational device, and for which the representation of nanotechnology was summarized into the "basics," delegated to experts invited to present nanotechnology to the public, or provided through the other components of the NISE project. The forums held at the Boston Museum of Science thus remained an isolated experiment, which conceived the deliberative device as a component of local policymaking. By contrast, the mainstream position of the NISE network made deliberation a way of "engaging" with the newly acquired knowledge, and making individual citizens reflect on how nanotechnology would affect them. As such, deliberation became a component of "informal science education," and a domain about which the NISE network could then propose expertise on.

The Democracies of Science Museums

In European and American science centers, defining and operating "new" and "more democratic" practices for the museum are shared concerns. But these new missions differ across the sites we examined, and these discussions are directly related to different problematizations of nanotechnology. Indeed, French, European, and American science centers have helped us illustrate three different roles for the science museums. The French case is that of the construction of a representational system, in which visitors actively participate in the display and practice of nanotechnology's various components (including the "public debate"). In the European case, the science museum is expected to represent nanotechnology and its social, ethical, and legal aspects, while paving the way for a "scientific understanding of the public" meant to replace "public understanding of science." Eventually, the American "informal science education" enacts a political model based on deliberation, for the sake of making an individual citizen knowledgeable about a field that will impact him or her, as a consumer, voter, or worker. The representations that are constructed by the science centers are tightly linked to nanotechnology policy, not less because of the funding links among science policy programs, research institutions, private companies, and science centers. They are not at-a-distance representations of a passive domain: they lead to the construction of material objects in the French case, they are connected to the making of science policy programs in the European case, and they produce nanotechnology's publics and concerns in the three examples. This chapter has stressed the importance of the representation of nanotechnology for its expected publics, as (if not more) for science policy officials. It also leads to the conclusion that the science center's position may vary, and in any case needs to be negotiated with many actors. But in all cases, nanotechnology programs involve science museums. In return, the display of nanotechnology in science museums participates in the problematization of nanotechnology as an entity gathering objects, futures, concerns, and publics. It makes it a matter of experiments with "public debate" in France, a problem of ELSA and of science policy options open for direction in Europe, and a question of understanding a stable scientific field in the United States. In this latter case, science museums emerge as a specific source of expertise about informal science education, expected to be separated from a field it displays. This agencement is based on technologies of representation and dialogue (among them the forum), which can possibly circulate from nanotechnology to other domains. Chapter 3 will discuss further this type of agencement, based on "technologies of democracy" expected to stabilize modes of democratic organization.

Situated within current interests in science communication for interactivity and the representation of science in the making, the examples discussed here also illustrate variations within these trends and the specific issues raised by nanotechnology. The agencements described in this chapter are all interactive, they all challenge the representation of scientific "facts," they are all meant to go beyond public instruction by innovating in the field of science communication. Concerns for "two-way dialogue" and "engagement" are explicit in all three cases, in which the democratic ambitions of science museums are visible. This should not be considered as the end point of the analysis, but as an invitation to look into the types of democracy that the museums produce, the nature of the representations on which they base it, the kind of people they aim to construct in order to fit with it, and their connections with wider institutional constructs. Indeed, the agencements described in this chapter are quite different, and they engage different democratic constructions, as exhibit designers, scientists, and anti-nanotechnology activists argue over the ways to shape public concerns in the science museum, and over the modalities of the involvement of publics in the development of nanotechnology. The French nanotechnology exhibit challenges the very idea of representation at a distance and proposes to integrate the visitor in the display and practice of nanotechnology, in the secluded place of the science museum. It proposes to experiment with the forms of public communication and public/private relationships, in order to make nanotechnology a matter of "public debate." European nanotechnology policy officials made the intervention of science centers a problem of democratic legitimacy by exploring the ways in which the "European public" can be heard. The problematization of nanotechnology as an issue of common values for the diverse European public draws a democratic space that is not characterized by electoral representations and constraining legal interventions, but by the mobilization of the European public through distributed dialogue processes in order to provide upstream elements for policy choices. Eventually, the choice for "informal science education" in the United States makes nanotechnology yet another scientific field for people to understand, possibly through deliberation. Problematizing the role of science museums within the American federal nanotechnology policy remobilizes well-known figures of the American polity, among which are the "informed" and the "deliberating" citizens (Manin 1997; Schudson 1998), who are expected to participate in the success of the development of the field.

The sites of problematization encountered in this chapter are connected with each other. Connections are drawn by the actors themselves, as they compare the initiatives undertaken elsewhere (as, for instance, the American museum experts do), or as they circulate, like the director of the Grenoble science center, from national to European science communication projects. The sites are not isolated from others outside of the science communication domain. French science museums attempt to answer the contestation voiced by anti-nanotechnology activists, while trying to involve public and private actors in the sponsoring of exhibits. American and European initiatives in nanotechnology communication or informal science education are directly linked with public policy choices about the responsible development of nanotechnology. This suggests pursuing the study of the problematizations of nanotechnology as they develop in other sites.

3 Replicating and Standardizing Technologies of Democracy

Technologies of Democracy for Nanotechnology

Chapter 2 describes sites where nanotechnology is represented in public, and, simultaneously, where publics are crafted for nanotechnology to be made a public matter. In the science museum, nanotechnology is problematized as a public issue deserving the engagement of various publics, in ways that contribute to shape the French, European, and American democratic spaces.

This chapter follows up on these examinations by considering agencements grounded on devices expected to make people "participate" in collective discussions about nanotechnology. These devices have been referred to in diverse ways: "participatory instruments," "dialogue mechanisms," "public engagement tools." Nanotechnology has been considered an important area for public engagement since it became a major domain of science policy in Europe and the United States. Nanotechnology proponents considered that public engagement was needed in order to ensure the success of the enterprise, while social scientists saw in nanotechnology a unique opportunity to renew the forms of integration of society into technological development (Guston and Sarewitz 2002; Macnaghten, Kearnes, and Wynne 2005). While the general topic of "public engagement in nanotechnology" is widespread, has been discussed in numerous government reports, academic publications, research projects; and has been translated into countless "dialogues," "debates," or "stakeholders' meetings," the practice of it and its overall objectives are diverse and may appear contradictory. Is the objective to convince society of the benefit of nanotechnology? Are people expected to actively participate in the material construction of nanotechnology objects?

Examining these questions requires two connected investigations. The first one deals with the mechanisms expected to produce "publics."

"Technologies of democracy" have recently been the focus of analytical attention (Felt and Fochler 2010; Lezaun and Soneryd 2007; Laurent 2011). Drawing from STS research on scientific instruments, these works have proposed to examine the sociotechnical apparatus necessary to produce "participating publics"—and, more generally, citizens playing their part in democratic life (see, for example, Rose 1999). Focusing on the instrumentation of participation is a way of describing in detail the experiments and demonstrations performed within participatory setting. Technologies of democracy constitute particular agencements based on the separation between an instrument expected to frame public discussions and define modes of action for participants on the one hand, and the particularities of issues expected to be discussed in a participatory manner on the other hand. These agencements are not neutral. They require bodies of expertise specialized in democratic practices, and the abilities to standardize these democratic practices.

Therefore, one can expect that problematizing nanotechnology as an issue of public participation is also problematizing democracy as a technological issue to be dealt with by appropriate expert knowledge. In order to display these problematizations, and the differences among them, one needs to develop the analysis along a second line of investigation, which interrogates the constitution of political spaces within which technologies of democracy have value. For instance, the construction of "mini-publics" in the United Kingdom takes place within a network of hybrid institutions having connections with consultancies and public bodies, which has eventually streamlined the forms of participatory democracy as devices expected to constitute small-sized neutral publics (Chilvers 2010). Technologies of democracy are experimented with, and for these experiments to have value, they require, as scientific experiments (Latour 1988b), a transformation of the spaces in which they circulate. This is particularly important to keep in mind for our concern here, since the would-be participatory publics are often imagined to be national. In two examples I discuss in this chapter, technologies of democracy are experimented with in order to produce national publics for nanotechnology crafted as a state-led science policy program. This directly raises questions connecting the manufacturing of national publics and the definition of the sources of democratic legitimacy. Comparative analysis is then a way to display different conditions for replicating technologies of democracy, different roles for expert knowledge to play in political practices, and, eventually, different problematizations of nanotechnology as a problem of public participation.

This chapter presents two types of problematization sites. The first one is the experimental site itself, where technologies of democracy are used, possibly replicated after previous cases. I contrast a national public debate on nanotechnology conducted in France in 2009–2010 and an American experiment with a nationwide citizen conference organized in 2008. In both cases, the replication of known technologies of democracy required managing the specificities of nanotechnology, and ensuring that the participatory device was properly separated from the issues being discussed. But the two cases also differ in that the attempts to construct national publics for nanotechnology framed as a national issue problematized both nanotechnology and democratic practices in different ways. While the French public debate made nanotechnology a topic for a state experiment with public participation, the American initiative was construed as a social scientific device through which the value of a technology assessment method could be demonstrated.

The second site of problematization that this chapter considers is international. Looking at the Organization for Economic Cooperation and Development (OECD), and particularly a project aiming to standardize methods for "public engagement with nanotechnology," the last section of this chapter examines the processes through which agencements based on technologies of democracy separated from the issues on which they are applied are stabilized. Such processes end up problematizing nanotechnology in ways that are acceptable within the international arena, and simultaneously restate the importance of the distinction between international expertise and sovereign national policy choices.

A State Experiment with Public Participation

A National Debate on Nanotechnology
The French Commission Nationale du Débat Public (CNDP; the National Commission for Public Debate), as noted in the prologue, conducted a nationwide debate on nanotechnology from October 2009 to February 2010. The CNDP is an "independent administrative authority," meaning that while it is a public body and a component of the French state, it acts independently from the government once it is commissioned on a particular project. The CNDP organizes public meetings based on the contribution of all interested actors, who are invited to write *cahiers d'acteurs* (stakeholders' brochures). Debates result in a report written by CNDP, which does not provide recommendations but presents the diversity of the arguments about the topic at stake. Founded in 1995, the CNDP mostly is mobilized

on local infrastructure projects,[1] and has developed an expertise about the organization of public debates in these cases. The CNDP has grounded its approach in the informal standardization of its debate methods, according to the legal requirements, which determine the duration of the debates but do not impose particular formats. The successive works of its members lead the commission to produce methodology booklets meant to inform new members of organizing committees about how to conduct public debates, while social scientists have been describing this as "a French experiment with public participation" and are sometimes members of organizing committees (Revel and al. 2007).

The public debates organized by CNDP do not target the statistical representation of publics. Rather, their objective is to ensure the representation of the various arguments of the groups with interest in the topic. Hence, organizers often invest a lot of time and energy to convince members of these stakeholder groups to participate.[2] Since 2002, the French government has commissioned CNDP to organize debates on "general options."[3] In 2009, seven French ministries commissioned CNDP to organize a "national public debate" on nanotechnology in order to "enlighten" public decision making on nanotechnology. This initiative was the direct consequence of a legal requirement, itself an outcome of a collaborative process that then newly elected President Sarkozy had initiated in 2007 in order to renew the French environmental regulation. The French government was thus legally bound to organize a national public consultation on nanotechnology, and decided to do so by commissioning CNDP to organize it.

For the French government, the national debate was a component of a more general approach toward nanotechnology, which consisted of a series of attempts at a "responsible" take on technological development. Some of these attempts targeted nanotechnology objects (chapter 4 will provide some illustrations of them), others, such as the national debate, were meant to engage nanotechnology publics. For CNDP members, the national debate was an opportunity to demonstrate the adaptability of the participatory device and its relevance for "general options" pertaining to large-scale science policy and environment regulation choices.

Both the French government and CNDP had strong interest in performing a convincing demonstration of the procedure's relevance for nanotechnology, of the state's commitment to the consensual development of nanotechnology. But replicating the CNDP procedure on nanotechnology proved difficult. First, contrary to infrastructure projects or to earlier examples of "general option" debates, the issue at stake was not localized in a particular geographical site. Second, the representation of nanotechnology, itself a global science policy program gathering objects, futures, concerns, and publics, is not self-evident, as chapter 2 demonstrated. The procedure was adapted in order to deal with these two characteristics. CNDP organizers planned meetings all over the country. Each meeting focused on topics linked to the industrial and research activities of the city where the debate was held, and, in some cases, developed a more general theme. For instance, the Orléans public meeting focused on the local nanotechnology industry (particularly cosmetics) and on consumer safety. This way, the organizers hoped both to ensure the national character of the public debate and restore the format that the CNDP was used to: it was supposed to ensure the focused and local representation of nanotechnology and public argument about it. But this careful choice was not sufficient to make the replication of the CNDP procedure on nanotechnology successful.

Representing Nanotechnology

As for all CNDP debates, the nanotechnology debate required that the commissioner produced a report presenting the issue, its technical characteristics, and the choices that were to be made about them. The report was an interesting piece as it was supposed to produce a single voice emanating from "the French government," describing what nanotechnology was and what decisions were to be taken about it. As seven ministries commissioned the debate, the report was a multiauthored piece. An advisor to the minister of environment was in charge of assembling the report. I met him as he was completing this task, which was a difficult one.[4] While the ministry for research insisted on the importance of nanotechnology for the scientific development of France, the ministry of health questioned the relevance of traditional risk–benefit analysis and called for the exploration of innovative methodologies for the public management of uncertainty. That different components of the state have different expectations and priorities on a given issue is not a novelty (and issues related to the protection of human health or the environment can be particularly contentious, for that matter). But what is more interesting in this case is that the French state submitted itself to a public demonstration of its ability to speak in one voice even before the majority of policy choices had been made. Nanotechnology thereby appeared as a political experiment about the ability of the state to display its commitment to consensual technological development, "consensual" in that it would gather all the concerned publics, as well as the various components of the state, whether they are in charge of health protection or industrial development support.

The French government eventually submitted a unique report authored by the seven ministries commissioning the debate, which attempted to cover the entire domain of nanotechnology, treating it along its different chapters, and in a somewhat self-contradictory manner, as a reservoir of new technological applications, as a science policy program deserving public support to be fully realized, or as a problematic issue because of uncertainty surrounding the potential health consequences of new nanotechnology objects.

The CNDP hoped to make this variety a suitable debate topic by dividing nanotechnology into themes related to the local particularities of the places where public meetings were held. But the topics chosen were regularly displaced during the public meetings, as participants raised questions not related to the foreseen topics. For instance, participants at the Orléans meeting discussed workers' safety or privacy issues related to the use of nanoelectronics, and others questioned the value of the overall French nanotechnology program. The processes through which participants in public meetings displace the predefined boundaries of the topics at stake have been described in detail by the political scientists interested in participatory formats, including those who have been working on CNDP (Jobert 1998; Fourniau 2007). During the nanotechnology national debate, these usual practices—necessary to question predefined framings, convince others to rally around a public cause—were multiplied by the uncertainty about the very nature of nanotechnology objects.

As the identity of nanotechnology substances and products remained unclear, participants did not agree, for instance, on the presence or absence of nanomaterials in food products. Consequently, the commission could mainly call, in the final report, for the "identification of substances" and for "information" about the uses of nanotechnology in consumer products,[5] without having been able to point to the various ways of defining these substances (see chapter 4). Hence, not only was it difficult to represent nanotechnology as a collection of separate "sub-issues," but also the very objects at stake (whether they were currently produced or foreseen for future developments) were of uncertain identities.

Representing Arguments/Publics

The questions raised by the representation of nanotechnology were further complicated by the challenges that the identification of the publics concerned by the sub-issues faced. While the logic of the organization of the debate was to constitute concerned publics for whom the topics discussed during meetings would make sense, the distinction among the sub-issues somewhat arbitrarily constituted by the organizers proved impossible to maintain.

Rather than the concerned publics that the organizers hoped to attract, the publics participating in the meetings were mostly of two natures. First, many participants attended the meetings in the hope of learning about nanotechnology. They were not members of the concerned publics the organizers had hoped to engage, but merely lay participants hoping to grasp some information about a new scientific field. Second, the most vocal participants were anti-nanotechnology activists, who considered that the national debate was part and parcel of a science policy program they opposed, and, as such, was to be fought against rather than participated in. At stake here was the separation between technologies of democracy and the issue on which they are applied. While the CNDP attempted to replicate an instrument that has been standardized (albeit in a flexible manner) independently from the issues at stake, the activists considered that the debate was a component of nanotechnology, which they understood as a global program of technological development in need for public support.

The most striking feature of the CNDP debate was indeed that contestation was extremely vocal. Meetings were repeatedly interrupted by opponents claiming that the debate was merely a trick meant to make the public accept an unquestioned program of development of nanotechnology. The organizers attempted to answer this opposition though several adaptations of the CNDP procedure. For a couple of meetings, they separated the meeting place into two rooms, a closed one in which invited experts were present, and a second one open to the general public. Two organizers were present in the open room to facilitate the discussions, and exchanges were then supposed to be possible by phone or on the Internet. As this did not diminish the contestation, the organizers eventually set up closed meetings in which the participants had to be identified (see the prologue). They also relied heavily on online contributions, which were far easier to control. This online meeting ground was where, in the terms used by the president of the CNDP, "real debates could happen."[6] But the elimination of this unwanted public was not enough to overcome the problem of the representation of the field of nanotechnology and the identification of its publics. Once purified from the publics that were "too engaged" (engaged to the point that they opposed the debate as such), the legitimacy of the debate was questioned by other actors. Civil society organizations withdrew from it, while the media attention was mostly turned to the "failure" of the CNDP to organize peaceful public meetings.

A State Experiment

The French government eventually replied to the national debate by raising several points, among them its commitment to soliciting the information about nanotechnology that it would provide to the French citizens, and to undertaking initiatives in order to make nanotechnology objects governable entities in their own right (see chapter 4).[7] As an experiment in nationwide public participation that was expected to enlighten public debate, the nanotechnology debate is considered a failure by the officials involved, who regularly complain in private conversation about the impossibility of conducting a debate in the context of such "radical opposition" (in their own terms). Other actors are more nuanced. In particular, people involved in the organization of the debate considered that the debate was a way of making visible the issue of the identity of nanotechnology objects.

The interesting point for our reflection here, however, is not to decipher whether or not the debate was successful. It is to consider it as a particular site of problematization based on the replication of a technology of democracy. The CNDP debate appears as a site where nanotechnology is problematized not only as an issue of publics (publics to be identified, public to engage, and radical publics to eliminate), but also as an issue directly related to the nature of the French state. With the nanotechnology debate, the state experimented with public participation as much as it was experimented with.

Indeed, the debate is situated within a series of evolutions by which the French public administration attempts to make the organization of participatory discussions a component of its centralized expertise, and local citizens concerned by controversial issues become actors in the French polity. Through the CNDP procedure, the French state committed itself to a real-scale test, whereby it brought to the fore the modalities of its action in public, its own possibility of speaking in one voice, and its involvement in the collective treatment of social and technical uncertainty. This experiment is then not just about the more or less successful replication of a given technology of democracy (although this is certainly part of the picture), but also about the ways in which the state may act in public to act on technological development. This is directly related to other initiatives undertaken by the French government, which will be described in the following chapters.

Testing Social Scientific Expertise on Public Deliberation

Producing a National Public for Nanotechnology in the United States

The French public debate on nanotechnology is an example of an attempt at producing a national public for nanotechnology. There is another notable one, in the United States, which was based on an American version of the consensus conference, whereby a small group of people is constituted, trained, and asked to provide an opinion about a topic they were initially unfamiliar with. This American example is interesting for a number of reasons. First, like the French nanotechnology debate, it is an attempt at producing a national public for nanotechnology. Second, it is an illustration of the replication of the consensus conference model in a format that was meant to produce social scientific demonstrations for the development of an American approach to deliberation. This makes this case quite different from other uses of the consensus conference format on nanotechnology. France is another country where the consensus conference was replicated on nanotechnology. The Ile-de-France regional council (that of the Paris regional administrative area) commissioned a consensus conference on nanotechnology in 2008. Before the national debate on nanotechnology, this was another test in new forms of policymaking. As we will see in the following pages, it provides a useful counterpoint to identify the specific dimensions of the American conference.

The U.S. initiative originated in the work of researchers including political scientist Patrick Hamlett at North Carolina State University, who had developed the "U.S. version of the Danish consensus conference"[8] called the Citizens' Technology Forum. A citizens' forum is organized as follows: a group of citizens is selected and its members receive background material to read before they first meet. They then work together, with a facilitation team, in order to prepare questions to be asked to "content experts." Using the answers they receive, they write recommendations about the issue under discussion. This was the process followed in 2008 when the National Citizens' Technology Forum (NCTF) was set up. Coordinated by Patrick Hamlett, the NCTF was meant to make nanotechnology a topic for deliberation at the national scale.

Hamlett's previous experiences with the Citizens' Technology Forum, particularly on topics related to biotechnology, had been opportunities for him to study "pathologies of deliberation," that is, the processes through which discussions are captured by certain participants or topics. In a 2003 paper, Hamlett had explained that the social scientist should locate these pathological processes in order to be able, at a later stage, to counter them

(Hamlett 2003). As another occurrence of the Citizens' Technology Forum format that Hamlett had developed, the NCTF was supposed to be an opportunity for social scientists to describe these processes, make sure that deliberation happened and was not captured by the most powerful actors, and eventually demonstrate that citizens *can* deliberate in a valuable way. This demonstration was addressed to two publics. First, academics were to witness the social scientific value of the experiment about deliberative practices—and numerous publications commented on the device.[9] Second, policymakers were to witness the value of deliberation to produce relevant public advice (Cobb and Hamlett 2008).

These targeted demonstrations make the NCTF a particular case. The French consensus conference on nanotechnology commissioned by the Ile-de-France regional council also had "citizen" in its name. It was a "citizen conference" (*conférence de citoyens*), as the other participatory exercises of this type conducted in France.[10] But what linked it with citizenship was not, as in the NCTF, the study of deliberation and the demonstration of its value. Rather, it was the connection with public bodies hoping to demonstrate their commitment to public participation that made the *conférence de citoyens* a matter of citizenship. The conference was entirely filmed, and was the topic of an educational DVD the regional council paid for, and which was then used to display to political science students the value of the citizen conference in particular, and of participatory democracy in general.

The French conference did not function as a research instrument. It was organized by IFOP, an opinion polling company that had made the organization of citizen conferences a component of its market offerings to public and private actors. When I met the person in charge of citizen conferences at IFOP, one year after the nanotechnology one, he told me that what mattered for his company was that it succeeded in demonstrating its ability to "make things work." This expression, meant, for him, that the conference was to produce what he called "a reasonable opinion"—precisely what IFOP clients were looking for, including the Ile-de-France regional council hoping to demonstrate the value of public participation.

The stakes were different at the NCTF. Beyond the plea for deliberation, the NCTF was also an opportunity to demonstrate the possibility for new technology assessment methodologies. It was part of a program funded by the National Science Foundation (NSF) after the 21st Century Nanotechnology Research and Development Act of 2003 had inscribed in federal law the need for research about the "ethical, social and legal implications of nanotechnology," and its integration within science policy programs (see chapter 1). The Center for Nanotechnology in Society (CNS) at Arizona State University received an NSF grant to conduct "real-time technology assessment," one of the components of which was "public engagement and deliberation" with nanotechnology issues (Barben et al. 2008; Guston and Sarewitz 2002; see chapter 5). The NCTF was part of the public engagement component of the program, which was expected to experiment with original ways of connecting the development of nanotechnology with its publics.

The NCTF was particularly interesting, for that matter, since it was organized in six sites that spanned across the entire United States. The nationwide character of the experiment was made possible by the use of "keyboard-to-keyboard" exchanges among the six sites. During the first and third weekends of the NCTF, panel members physically met in each of the six sites. During the second weekend, local groups exchanged comments with each other through an online process, and groups were formed that included members from each of the sites. Group members were supposed to talk together online, one group at a time while the others watched the ongoing conversations. It was through the Internet part of the discussion that the NCTF could become a "truly national" event, and an innovation in the practice of the Citizens' Technology Forum.

Producing Content to Deliberate About

There were two necessary conditions for NCTF to realize the dual demonstration of the value of deliberation and the feasibility of a participatory form of technology assessment. First, the technical content of nanotechnology had to be stabilized enough for deliberation practices to be the sole topic of examination. For instance, deliberation specialists evaluate the quality of deliberation in Citizens' Forums using the "internal political efficacy" (IPE), which measures the acquired knowledge of the participants as well as their confidence in it and in their ability to use it publicly (Cobb and Hamlett 2008):[11] participants are asked to answer a series of questions about the forum topic and grade their confidence in their answers. The measure of IPE defines the value of deliberation in terms of learning, awareness of the knowledge gain, and ability to use it to act as a knowledgeable citizen. It draws a boundary between what is known and unproblematic (the issue itself) and what is being done in the procedure (the transformation of participants into knowledgeable citizens).

Second, as the NCTF was also expected to demonstrate to policymakers the value of a renewed approach to technology assessment for this emerging technological domain, nanotechnology could not be a mere pretext for the deliberation of citizens, but the organizers had to prove that it could be acted upon in all its complexity.

This tension was dealt with through the choice of "human enhancement" as a relevant topic for the deliberative exercise. Human enhancement gathers all the technologies that are designed to "enhance human performances." These technologies (e.g., brain stimulation probes) are transformed by nanotechnology, especially as it converges with other technological domains. As a significant area of converging technologies, human enhancement was considered appropriate since it allowed participants to discuss existing technologies, future prospects, and the "societal implications" of nanotechnology.

In choosing human enhancement, what mattered was that the domain connected the various dimensions of nanotechnology and the possibility of representation through scenarios describing the potential evolution of nanotechnology. This choice was not as problematic as the definition of specific topics for the local meetings of the French national nanotechnology debate. Nor was it as contentious as the choice of the Ile-de-France conference to define nanotechnology as a set of industrial applications, which, for many members of the supervising committee, was not enough to represent the potential future evolutions of the field (the social scientists sitting in the committee opposed IFOP on that point). By contrast, the NCTF organizers could describe the future of nanotechnology development within the safe perimeter of the social scientific laboratory through scenarios, and isolate the content presented to panels to the deliberative practices that were to be looked at. Through the use of these scenarios, the subset of nanotechnology that was chosen for its paradigmatic representation of the field could then be appropriately presented to the participants, and the future could be deliberated about. The boundary between the information to be provided and the deliberation to be conducted, and then studied, could be effectively maintained.

Turning Panel Members into Deliberating Citizens

Online exchanges made it possible for the NCTF project to be construed as a national deliberation. The Internet has another virtue. It could be used to ensure that the deliberating citizen was not captured by special interests. As the Citizens' Forum was meant to evaluate and correct "pathologies of deliberation," having participants interact on their screens was conceived as a way of purifying them from visual and oral signals, and thereby limiting "preconceptions and stereotypes" (Prosseda 2002, 220).[12] The Internet was also a powerful tool for control of the issue being discussed.

The software used for the NCTF allowed the organizers to disconnect some people, thus controlling who could speak and exchange with the content experts who were supposed to answer the questions raised by the participants. The "truly national" dialogue could not happen without fine technical arrangements about who could speak with whom. While one group gathering participants from each of the sites was active, the other participants were expected to watch the screen and read the exchanges. That way, the organizers expected "real deliberation at the national level" (Cobb and Hamlett 2008) to happen. "Real deliberation" would involve a limited number of people each time, so that the moderator of the Internet session could make sure that every member of the active group had a chance to participate, and that the issue being discussed remained within the topic of "human enhancement." As the moderators had priority in the posting of messages, they could intervene quickly when they felt that questions were "too vague" or that they "did not really fall into the topic of human enhancement" (Cobb and Hamlett 2008).[13] For instance, as some people were trying to raise questions about nanotechnology-related health issues, they were quickly reminded that the topic was human enhancement, and that toxicological risk issues did not fall into that category. This might have been at the price of the dwindling interest of participants to get involved in the discussions.[14] But as many factors could destabilize the procedure (e.g., participants switching discussion topics, or intervening when they were not supposed to), online exchanges ensured that deliberation remained a stabilized topic of study, and that "normal" participants were indeed deliberating.[15]

For deliberation to happen within the NCTF, as elsewhere, techniques of discipline and control were required. Producing a deliberating citizen was also a concern of the Ile-de-France conference. But in this case, the main objective was not to make deliberation a topic of study, but to train panel members so that the device could indeed produce the "reasonable opinion" it was expected to deliver. This meant that the selection of panel members and the management of panel discussions by the IFOP *animateur* (moderator) were crucial processes.[16] They were also imperfect. Several of the IFOP organizers told me during interviews that one of the panel members proved difficult to discipline. He was, according to the facilitator, "far too critical," in that he was reluctant to discuss the risks and benefits of nanotechnology's industrial applications but questioned the overall need of developing nanotechnology in the first place. It took all the moderator's expertise about group management to contain this critical stance, and make sure it did not threaten the production of the "reasonable opinion."

Alternative Agencements

Evaluated from according to their self-defined objectives, both the NCTF and the Ile-de-France conferences were successful. They both produced satisfactory end products. Both the opinions about human enhancement that came out of the NCTF and the report on nanotechnology that the IFOP panel members wrote were well received. They all adopted the language of the needed balance between risks and benefits, and called for additional inquiry into the potential concerns related to nanotechnology, whether in ethical or safety domains. As such, they fitted well within the general construction of nanotechnology as a program expected to associate objects, futures, publics, and concerns and meant to be developed in a responsible way (see chapter 1). But, perhaps more importantly, both the NCTF and the IFOP organizers performed the expected demonstrations. The NCTF operated as a social scientific experiment, at the laboratory scale, in order to demonstrate to academics and policymakers the value of deliberation to create "informed public opinion" (Cobb 2011), and, more generally, the value of a participatory technology assessment that could contribute to "anticipatory governance" (Guston 2011, see chap. 5). The Ile-de-France conference produced a visible proof that a regional council could ask lay citizens to provide an informed opinion about nanotechnology in particular, and technological issues in general.

The price to pay was the elimination of alternatives. Attempts at reformulating the problem of nanotechnology in the terms of the anti-technology activists who later interrupted several meetings of the CNDP national debate could not be accepted within the Ile-de-France conference. It had to be inscribed in the policy landscape of the regional council and therefore could not radically question the interest of technological development. Accordingly, panel members who threatened to be "too critical" have to be eliminated, possibly at the selection phase, and discussion contents have to be carefully monitored. IFOP's credibility as an expert provider of the conference construed as an instrument for the production of "balanced public opinion" relies on the ability to perform these tasks.

The replication of the NCTF citizen forum was far less controversial than the French CNDP debate on nanotechnology, and less contentious than the Ile-de-France conference. But alternate objectives were proposed. Some of the organizers of NCTF, who were researchers at the University of Wisconsin at Madison, had been involved in another consensus conference two years earlier. After this previous conference, panel members had created a citizen group active in the public debate about nanotechnology regulation in the United States. When the National Nanotechnology

Coordination Office (a federal body in charge of the coordination of nano-technology federal activities) convened its first meeting on environmental, health and safety Issues related to nanoparticles, the group submitted written comments and one of its members flew to Washington for this meeting.[17] The organizers of the 2005 conference, some of whom were known scholars in the field of the sociology of social movements, had been helping the group organizing local debates after the initial consensus conference and identifying relevant partners for social mobilization about nanotechnology regulation. They saw the "democratic virtue" of the consensus conference in the empowerment of citizens it rendered possible (Powell and Kleinman 2008). By contrast, they considered that helping the panel members to engage on nanotechnology after the conference itself was "clearly not the main concern of the NCTF," which, in comparison with the Madison event, was "a little bit disappointing."[18] One of them contrasted the NCTF with the 2005 Madison conference in which "the framing was different," in that the organizers insisted from the start that the conference was expected to impact public decision making.[19] As a result, the recruited people were "concerned about the topic" and, as such, would "probably not have made it into the NCTF" (because of their involvement with the issue being too high).[20]

This episode is interesting because it displays the variety in the replication of a technology of democracy. The laboratory setting that the NCTF was meant to be was not the only possible use of this technology of democracy. The NCTF way of producing engaged publics was clearly not what the Madison people expected from a consensus conference. For them, that nanotechnology was still in construction implied that citizens needed to be engaged under the format of the social mobilization. In this latter case, the technology of democracy was not entirely independent from the particularities of nanotechnology: it had to be adapted in order to ensure a form of engagement targeted to the regulatory issues at stake. By contrast, the NCTF was conceived as a machinery of *not* engaging publics too deeply, in order to ensure an experimental intervention targeted at deliberation rather than nanotechnology. The mechanic at play in the Ile-de-France conference was similar, as IFOP's expertise in the production of "reasonable opinions" consists in ensuring what the institute defines as the "neutrality" of the panel, which is not supposed to be anything else than "moderate."

Replicating Technologies of Democracy

The agencements that we encountered so far in this chapter are based on technologies of democracy meant to be independent from nanotechnology, and replicated on this domain after having been used elsewhere. This particular configuration is "experimental" in that it is based on the demonstrative replication of technologies of democracy. But the experiments take various formats, and the demonstrations they perform are of different nature. While the replication of the CNDP procedure is a real-scale experiment that engages the French state itself, the NCTF is meant to be a laboratory experiment about deliberation, and the Ile-de-France *conférence de citoyens* a local test conducted by the Paris regional council. In all these cases, the value of the experiment is measured according to the integration of the singular initiative into broader political spaces, which it contributes to extend. The French nanotechnology debate and the Ile-de-France conference can be seen as attempts at extending the space within which the "French experiment with public participation" (Revel and al. 2007) is conducted, while the NCTF is inscribed within a research program on deliberation based on the Citizens' Technology Forum. Yet the replication of these technologies of democracy also offers opportunities for alternative understandings—whether technologies of democracy are constructed as objects of a radical critique (in the French case), or expected to directly contribute to social mobilization (in the American one).

Agencements based on technologies of democracy are interesting for a number of reasons. First, they problematize nanotechnology as an issue related to its publics: they aim to engage some of them while disengaging others, and seek to produce concerns about technological developments that are yet to be realized. They attempt to stabilize the concerns related to nanotechnology for them to be made topics of collective discussions—and the elusiveness of nanotechnology as a global program makes this objective a complicated one. Second, they provide instrumented devices meant to organize democratic life. As they make nanotechnology a problem of engaging publics, they also make democracy a problem of tailoring the right technology to do so. They are part of a wider movement making democratic life the topic of a specific expertise, whose value pertains to situated institutional constructs. Thus, the French expertise in public participation is tightly connected to the role of the state in technological development and the management of its associated issues, while the American expertise performing the social scientific experiment in the secluded space of the laboratory is designed as a demonstration addressed to public bodies seeking neutral expert advice.

Experiments with technologies of democracy such as these are sites where nanotechnology and democracy are jointly problematized. From there, one can extend the analysis to other sites. From the CNDP debate, one can attempt to look at other attempts of the French state to extend the perimeter of its expert competency, for instance, to nanotechnology objects. From the NCTF, one can start looking at other activities in the public demonstration of a renewed technology assessment. These explorations will be undertaken in, respectively, chapters 4 and 5. But before that, it is also useful to reflect on the particularities of the agencements based on technologies of democracy. They locate the problem of the democratic organization in the technical details of the instruments meant to produce participating publics. They distinguish between matters related to the production of publics and issues related to the technical contents of the topic at stake. These operations are not neutral, as the oppositions to the CNDP debate and the counter interpretations of the NCTF's value demonstrate. They contribute to "technologize democracy"—as critics of the biotechnology participatory attempts put it (Levidow 1998)—in different ways according to the institutional constructs they are part of and contribute to shape. Thus, CNDP as an independent administrative authority marks an attempt by the French state to integrate in the scope of its public expertise the ability to organize participatory discussions on controversial topics, while the NCTF signals the extension of the network of a social scientific expertise about technology forums. The common features of their technology of democracy formats, and the differences pertaining to national particularities compel us to examine sites where these commonalities and variations are analyzed, and potentially remade. The next section analyzes one of these sites, which made public engagement in nanotechnology a topic for international reflection.

Making Nanotechnology an International Problem about Publics

In the previous section, we encountered agencements meant to produce publics for nanotechnology, and based on technologies of democracy separated from the issues at stake. These agencements are valued within particular political configurations that they contribute to shape, whether it is the extension of state expertise on participatory matters in France or the importance of social scientific objectivity in the United States. The technology of democracy format is one particular modality of democratic life. One of the sites where it gets stabilized, particularly relevant in the case of the global regulation of technological innovation, is international expertise. This section looks at the OECD, an intergovernmental body of Western countries, and discusses some of its initiatives regarding the "engagement" of nanotechnology's publics. These initiatives are situated within a broader concern for the construction of a global market for nanotechnology, which was voiced early in the formulation of national nanotechnology programs. In these calls, GMOs were counterexamples, in that they were described as a case of failed regulatory harmonization of technical objects and market demand, opposing Europe and the United States.[21] The transatlantic divide was construed as a central risk to avoid in the development of nanotechnology, and avoiding it required international coordination in technical and social harmonization. The OECD was one of the places where these questions were asked, and, as the section will commence to explore, where they were framed by separating issues regarding technical objects from issues regarding publics to engage.

Public Engagement as a Set of Activities

One of the initiatives undertaken at the OECD about nanotechnology was the "Working Party on Nanotechnology" (WPN), created in 2007, which aimed to gather information about nanotechnology public policies. The WPN launched a series of projects, one of which was devoted to "public engagement in nanotechnology." In June 2012, the OECD released a report that originated from the work of the WPN public engagement project. The report was entitled *Planning Guide for Public Engagement and Outreach in Nanotechnology. Key Points for Consideration When Planning Public Engagement Activities in Nanotechnology* (OECD 2012).

The "key points" were illustrated by examples, which were public engagement mechanisms organized in OECD member countries. They comprised considerations related to the definition of the objectives of public engagement, or to the identification of relevant participants. Taking this report as a starting point, my interest here is to explore what the whole process says about the international ordering process, how nanotechnology is problematized, how publics are imagined, and how international government takes shape.

The starting point of the report was the affirmation of the central role of "the public and society at large" in the development of nanotechnology and for the "acceptance of the technology in marketable products" (OECD 2012, 3). This required that the various ways of communicating nanotechnology and engaging "wide audience in the debate" were examined as part of international cooperation: "Strategies for outreach and public engagement in nanotechnology have been identified as crucial elements of government policies regarding nanotechnology. The need to clarify how to communicate, with whom and how to engage a wide audience in the debate on nanotechnology, and in the development of policies related to it, has been a major point of discussion amongst policy makers" (ibid., 3).

The report made nanotechnology a problem related to publics to engage and communicate with for the sake of technological development. This problematization had two important consequences. First, it made "public engagement" a specific field of concern for the policy expertise exercised by the WPN. Second, it supposed that the problem of nanotechnology's publics was to be dealt with through planning and organizing "activities." The list of these activities comprised a variety of mechanisms, including "public lectures," "consensus conferences," "debates," "public hearings," "games, internet/web-based activity, blogs," "science festivals, science cafés, science weeks," and "science and technology museums, interactive science centres." The diversity of activities was reflected in the diversity of "objectives," as they appeared through the self-reporting of member countries. These objectives comprised "increasing public awareness about nanotechnology and its benefits and risks," "improving knowledge about ethical and societal issues," "initiating dialogue between stakeholders," and "enabling an informed public debate."

The wide range of public engagement activities was gathered under the following definition in the report:

For the purposes of this work by WPN, public engagement is a process that is:

Deliberative—emphasizing mutual learning and dialogue.

Inclusive—involving a wide range of citizens and groups whose views would not otherwise have a direct bearing on policy deliberation.

Substantive—with topics that are related to technical issues, and appropriate to exchange;

Consequential—making a material difference to the governance of nanotechnologies.

The 2012 OECD report makes public engagement in nanotechnology a crucial concern for policymakers, without defining in stricter terms than the preceding definition what public engagement could and should be.

One could easily question the similarity of the activities gathered under the sole heading of "public engagement," and argue that each of them is deeply rooted in national institutional constructs that made them poorly comparable with one another. Yet my interest here is less to evaluate the efficiency of the OECD approach than to analyze what it says about the problematization of nanotechnology in international settings. In the 2012 report, public engagement was based on a linear approach, in which objectives, mechanisms, outcomes, and evaluation were connected in a single reality, separated from the particularities of political contexts. The problem of nanotechnology's public was to be dealt with by picking and choosing the adequate "activity" according to the type of "audience" it was intended to target. This is the model of the technology of democracy expected to travel from one issue to another, as the American Citizens' Technology Forum or the French CNDP debate do. In the report, this meant that public engagement was a domain separated from the technical details of nanotechnology itself. Accordingly, "nanotechnology" was never discussed as an entity open for redefinition or transformation in the 2012 report.

This way of defining the problem of nanotechnology's publics can be seen as another case of what has been examined in the previous section. Both the French public debate and the American National Citizens' Technology Forum attempted to identify publics to engage in discussions about nanotechnology. But the specificity of the OECD report is that it is the outcome of international expert work. As such, making public engagement a matter of selecting activities without examining the content of nanotechnology should tell something about processes of international ordering. The question, then, is not about the "effects" of the report, but about the mode of reasoning that ends up framing it as it is. Questions that matter to us here include: Why does the OECD think like this? How does it produce knowledge about ways of engaging publics? How and why is the agencement based on technologies of democracy separated from nanotechnology stabilized in the international arena?

If one is to know why and how the OECD frames the issue of public engagement in the terms outlined in the 2012 report, then one needs to describe the working process of the Working Party on Nanotechnology. To explore in further detail what it means to make nanotechnology a problem related to the engagement of publics at the OECD, one needs to delve into the machinery of expertise production. How does one follow the successive stages of reports? How are they negotiated? What is eliminated? To answer these questions, I use ethnographic material based on an eight-month participant-observation period at the OECD, during which I worked at the WPN. This, of course, raises additional issues related to the role of the social scientist and the nature of his or her own engagement, to which I will return in chapter 7. At this stage, it suffices to say that the inside view of the participant observer is a lens (and sometimes not a passive one) for the description of the international work of expertise production.

The organization of the WPN follows that of all OECD working parties. The working party is run by a bureau composed of delegates of the most involved countries. Plenary meetings occur at regular intervals. They gather members of the OECD secretariat, and delegates from member countries active in the working party. Countries may send one or several people to participate in the working party. In November 2008, the email list of the WPN delegates comprised about a hundred names (mostly science policy administrative officials). WPN plenary meetings usually gather about forty people from about fifteen member countries. Each project is run by a steering group composed of a subset of the delegates involved in the working party, as well as members of the secretariat, who regularly meet, physically or by teleconferences. Projects may mobilize external experts, especially through workshops hosted by steering group member countries. They are presented and discussed during plenary meetings.

The 2012 OECD report is the outcome of a process that started in 2007, as the Working Party on Nanotechnology of the OECD Committee for Science and Technology Policy (CSTP) was created. The WPN launched a project that was supposed to examine the initiatives of member countries in the fields of "communications" and "engagement" about nanotechnology. It proceeded by gathering information about the work of member countries in these fields, in order to produce, in a later stage, the "points for consideration" that would make the core of the 2012 report. This two-stage process is interesting in that it reveals the micromechanisms at play within the OECD for the production of international consensus, and which ended up defining the problem of public engagement as that of the organization of "activities" about publics, separated from the examination of nanotechnology issues.

Collecting Information for International Purposes

Gathering information about public engagement in member countries required formalizing what "public engagement in nanotechnology" could be. The initial version of the questionnaire asked member countries to explain the ways in which public engagement had "influenced policies related to nanotechnology." The authors of this first version were members of the British delegation, who were familiar with the growing interest in the UK for "upstream public engagement" in nanotechnology (Wilsdon and Willis 2004), and a French legal scholar working at the OECD secretariat for the WPN. Their objective was to situate member countries on a scale according to their level of "public participation" in nanotechnology policymaking—direct involvement in regulation making or in the choice of research priorities for public research being at the top of the scale. Writing the questionnaire then took several months, as delegates from member countries reworked the initial proposal. Two elements were transformed. First, the explicit identification of the objective of public engagement with public participation in policymaking was displaced, so that the objective of public engagement became open for member countries to decide. In fact, the final version of the questionnaire asked them to list their objectives, which, as the 2012 OECD report shows, were diverse. Korea, for instance, could list among public engagement activities public communication initiatives aimed to turn high school students into potential university students in nanotechnology, while the UK centered its contribution on its objective of "upstream public engagement" expected to involve various publics at an early stage in policymaking. Eventually, public engagement was enlarged to comprise all activities targeted to nonspecialist publics, whatever their objectives were.

Second, the general description of public engagement as an involvement of nonspecialists in nanotechnology had to be refined. This took the form of the introduction of several questions about "audiences" to which public engagement was expected to be addressed. As processed through devices such as description tables and examples provided to help delegates fill up the questionnaire, public engagement in nanotechnology could then appear as a collection of "activities" characterized by target audiences, various modalities of planning, and expected outcomes. The questionnaire had to leave enough room in the definition of public engagement for all members of the steering groups, and, more generally, of the WPN to participate in the questionnaire study, and thus be recognized as active players in the field of public engagement in nanotechnology. Focusing on activities allowed the WPN to produce knowledge about public engagement that did not threaten to evaluate the effectiveness of this or that national approach. It made it possible to reach international consensus by ensuring a separation between the production of international expertise about public engagement and national policy choices about the objectives and means of public engagement. Policy expertise was not supposed to cross the link between the international expertise and the national sovereignty decisions, since the international cooperation also meant that countries with very different democratic organizations cooperate.

Writing Like the OECD

Gathering information was only the first step of the project. "Best practices" were then supposed to be identified in order to produce guidelines about how best to engage the public in nanotechnology. During the April 2008 WPN plenary meeting, the definition of "public engagement" to mean "deliberative," "inclusive," "substantive," and "consequential" was chosen, as a way of gathering the whole range of "activities" identified in the questionnaire phase. Introduced by a British STS scholar and meant to reflect the perspective of upstream[22] public engagement, this definition was used to characterize all public engagement activities.

The crafting of the guidelines occurred during a project workshop that was organized in October 2008 in Delft, in the Netherlands. My first participation to the work of the WPN took place there. I was invited to speak as an expert sent by the French delegation. The objective was to reflect on the initial results from the questionnaire study, start working on the report of the public engagement project, and elaborate preliminary guidelines that would then be refined by the OECD secretariat and the steering group members to become the "Points for Consideration." Presentations were given by country delegates or invited experts, and reflected the diversity of the national experiences as reported through the questionnaire. During one of these presentations, I discussed some examples of public dialogue undertaken in France, and talked about the Ile-de-France *conférence de citoyens*.

For all their differences, the various perspectives could all be said to be "deliberative," "inclusive," "substantial," and "consequential." For instance, public perception studies were to be conducted through "dialogues" involving "a wide range of participants" in discussions about "appropriate topics." This work was expected to inform "communication and dialogue strategies." Hence, public engagement as defined as a measure of the public perception of nanotechnology could be said to be "deliberative, inclusive, substantive, and consequential" as the WPN definition contended. This is of course a different understanding of public engagement than, for instance, the French national nanotechnology debate, or the American NCTF. Yet the WPN definition could be used to encompass this variety.

Later in the Delft meeting, the WPN definition served as a general framework for the discussion among experts from the national delegations and OECD staff members. I participated in this discussion, during which participants attempted to formalize the variety of national experiences that had been collected in the questionnaire and presented earlier during the workshop. They did so by bringing together bits of expert advice from the workshop, information gathered from questionnaires, and their personal experiences as country delegates involved, in one or another, in various engagement activities. For me, as a soon-to-be internal observer, the most striking aspect of the discussion at that time was the combination of the minimal formalization of the participants' contributions and the extreme carefulness of the team from the OECD secretariat when writing "general lessons" that would later become the "key points" of the 2012 report. The lessons were not supposed to imply, even tacitly, that one version of "public engagement" was better than the other, but they also needed to bring together the diversity of countries' experiences into a coherent whole.

This resulted in a dual version of public engagement. On the one hand, the heterogeneity of what would eventually figure in the 2012 report started to appear as lists of "activities," "objectives," and "outcomes" at that time. On the other hand, the objectives of public engagement in nanotechnology appeared to be about the publics and *only* them, as the definitions of successive "points" focused on the practical organization of "activities"—a way of not entering difficult negotiations about what the problems were related to nanotechnology objects and programs.

Restabilizing the International Organization
The initiatives undertaken at the WPN ended up defining the problem of nanotechnology's publics as that of the expertise on technologies of democracy distinguished from the technical particularities of nanotechnology. This reflected the repartition of work between different components of the OECD. At the OECD, there was another working party specialized in nanotechnology, the Working Party on Manufactured Nanomaterials (WPMN). The separation of work between the two working parties was to be carefully maintained, and nowhere is this more visible than in situations where the boundary was threatened.

At the November 2008 WPN plenary meeting, Austria proposed to host a roundtable that would aim to identify "governance frameworks" for nanotechnology. The link with the public engagement project was clear for the member of the Austrian delegation who proposed to organize the roundtable. For some members of the national delegations, such an initiative appeared as an opportunity to reflect on "new governance models." The French delegation, for instance, repeatedly insisted on the need to push for the integration of publics' perspectives in nanotechnology policymaking. Members of this delegation were very much in favor of initiatives that connected the expertise about public engagement and the expertise about risks, and backed the Austrian proposition. At that time, they were starting to think about the future national public debate, and were already engaged in "redefining the conditions of state expertise to take uncertainty into account"—as one of them said to me during an interview.[23]

The organization of the roundtable was to be done by the Austrian Technology Assessment Institute and the WPN. As I was interested in the topic, I was involved in the organization of the roundtable for the WPN, and thus worked with the Austrian delegation to refine the agenda.

Following a suggestion from my part, the focus of the roundtable was defined as "policymaking in uncertainty."[24] The draft agenda proposed "parallel sessions" on "policy instruments for dealing with nanotechnology risks," namely "codes of conduct," "voluntary measures for the industry," and "participatory models and inclusion of lay people in regulatory processes." The example of a specific nanoparticle ("possibly nano-silver") was to be considered to provide illustrations of "risk governance in context of uncertainty."

The agenda was not satisfactory for the WPN, because of the repartition of work between the WPN and the WPMN. Hence, a distinction that my colleagues at the OECD were concerned about, and that the risk governance roundtable was on the verge of ignoring: "WPMN does risks, and we do policy." This was the phrase of a senior staff member who gave his criticisms of the draft agenda quoted earlier. For him, the initial proposition considered "risks and not benefits" and mixed up "science and policy."[25] I was then summoned to a meeting with him, during which he explained: "The mandate is clear: WPN does policy. We develop policy and benchmarks that ensure the responsible development of nanotechnology. WPMN does technical work. It asks whether the regulatory system is functioning for nanotechnology."

Therefore, any hint that nanotechnology risks would be looked at during the risk governance roundtable would be suspicious. It would threaten to shake the institutional repartition of work, and bring the people of the OECD secretariat in charge of the WPN on the verge of going beyond their mandate. What was to be done then? As the senior staff member put it to me: "You can't do a meeting with nanotech risks. What you can do is governance. What are we trying to do? What are the governance tools?"

Hence, the solution: as "policy instruments in uncertainty" threatened to cross the line between technical examinations of risks and work on policy options, "governance" would be an appropriate framework. Consequently, the WPN roundtable was eventually organized as a workshop on "communicating knowledge—communicating uncertainty,"[26] which examined "the path from risk assessment to risk management" in the first parallel session. "Participatory processes" and "voluntary measures" were still topics for discussion in two other sessions, yet on condition that "it (was) not nanotech risks that were talked about." Consequently, neither the "participatory processes" nor the "voluntary measures" to be examined would potentially intervene in the definition of nanotechnology risks.

This episode is less interesting for its anecdotal value than for what it says as a breaching experiment, as Harold Garfinkel would have said, rendering visible what was otherwise so much inscribed in everyday work practice that it did not have to be made explicit. Thus, the allocation of work between WPN and WPMN was a way of distinguishing the technical expertise about risks and the policy expertise about public engagement. This separation was seen as a condition for the international organization to function in acceptable ways. It was far from neutral in terms of the problematization of nanotechnology it made possible—rendering it impossible to conceive otherwise than in disconnecting the problems of nanotechnology's publics from those of nanotechnology objects.

There were other situations where similar breaching experiments forced OECD staff to restabilize this separation. Thus, a member of the French delegation proposed, during a WPMN plenary meeting I attended, to inquire into "the possibility of a governance framework for nanomaterials risk prevention" and consider the "integration of stakeholders."[27] The proposal did not receive any approval. Indeed, it appeared to be "policy expertise," and, as such, fell "within the area of expertise of the WPN" as stated later by the secretariat. French actors multiplied the propositions within WPN and WPMN that threatened to displace the science/policy boundary on which the work of the international organization was based: they were constantly rejected by the secretariat. Eventually, nanotechnology expertise at the OECD needed to be demarcated as "technical" and "policy" related to ensure that the organization could indeed produce it. Attempts to blur this demarcation by delegates (such as the WPN Austrian delegate, or the French WPMN ones) or misbehaving members of the secretariat (like myself) thus implied additional work to make sure that it is maintained, and that delegates and staff members behave properly.

An International Mode of Reasoning

The OECD is a site where nanotechnology was problematized as an issue of different, and separated components. "Publics" were to be examined separately from nanotechnology objects. And issues related to nanotechnology's publics translated into a series of initiatives making "public engagement" a matter of technologies of democracy, further separating an expertise about "publics" and how to engage them from an expertise about nanotechnology objects. It is through the description of the working processes of the OECD that one may identify an international mode of reasoning at play in establishing these separations. What is at stake is the possibility to ensure the smooth production of expert knowledge separated from the sovereign choices of the member countries, and the possibility to ensure a demand for the global nanotechnology market. Eventually, the outcome of this expert knowledge is a restabilization of a format of public engagement based on the circulation of technologies of democracy separated from the issues on which they are expected to be applied.

The OECD categorizes its intervention in the development of international cooperation on nanotechnology by separating the problem of nanotechnology's publics from the problem of nanotechnology's objects. This boundary work is at the heart of the institutional organization of the international institution and a crucial feature of its activities. Accordingly, this chapter displayed the work needed to ensure that only the publics of nanotechnology are discussed. Yet, there are international sites where objects are explicitly discussed. Chapter 4 pursues the description of the international problematization of nanotechnology based on the distinction between international expertise and sovereign political decisions. It uses this analysis as a starting point, and further contrasts it with other examples in France, Europe, and the United States, all related to the government of nanotechnology's elusive objects.

Governing

4 Making Regulatory Categories

How to Define Nanomaterials?

In 2009, I met Daniel Bernard for an interview. He was the head of a research unit of a major French chemical company, which set up a carbon nanotubes production unit (the largest in France) in the southwest of the country. Back then, Bernard used to say that the nature of nanotubes development within his company was "empirical." To him, "empirical" meant that producing nanotubes was an entirely experimental process, neither modeled nor controlled except for the examination of physical and chemical properties of output products. In other words, it was only after much feedback from its customers that the production unit managed to manufacture "good" nanotubes, fit for the required uses (in that case, building materials). Bernard paid particular attention to this empirical process because even though it met customers' requirements, it also was at the heart of the nanotubes identification problem. What made these freshly manufactured nanotubes "brand new"? Could they be patented? Should they be registered with the administration as a "new" chemical with specific hazards? Was each type of manufactured nanotube—whether it differed in size, diameter, flexibility, rigidity, or number of walls—to be distinguished from others within the national, European, and American regulations with which Bernard's company had to comply?

This chapter focuses on the practical problem of identifying nanomaterials through categories. By delving into the construction of regulatory categories, I aim to question the processes through which "new" chemicals come into being, for companies and public bodies attempting to control them. This requires locating the sites where the "nano-ness" of materials is discussed. This chapter argues that these sites are crucial places for our more general inquiry into the joint problematizations of nanotechnology and democracy.

Science policy programs do not offer a clear definition of nanomaterials. They use a size criterion to define nanotechnology, usually in approximate terms—the 100 nm upper size limit is often used. Given the main characteristic of nanomaterials is the smallness of their components, size seems to be an obvious criterion. Using a size criterion would require drawing a boundary between what is bigger or smaller than "nano." But the issue is not that simple. If the definition of nanomaterials is an attempt to restrict the production or the use of supposedly hazardous chemicals, then this definition needs to use criteria enabling a good identification of hazards for each substance. And size itself appears rather limited. As many scientists remarked during the early years of nanotechnology development, what determines many of the properties of small-sized objects, including their toxicological property, is their surface, where chemical reactions happen. Thus, it would make little sense to identify potentially hazardous substances using a size criterion while the surface criteria seem more closely related to the hazard profile.

In all cases, choosing a particular criterion demands technical and regulatory infrastructures. It is at this point that the articulation between objects, futures, concerns, and publics at the core of nanotechnology plays a distinctive role. Nanotechnology programs are developed for the sake of scientific and market development, keeping in mind the objective of anticipating technical and social risks. For the proponents of nanotechnology programs, this requires public intervention in order to make sure that nanotechnology objects are developed in a way that will not cause public controversies.

This is an acute issue in a situation where the use of nanotechnology objects in industrial products and processes has arguably skyrocketed. An inventory widely circulated is that of the Project on Emerging Nanotechnologies (PEN), conducted at the Washington, DC-based Woodrow Wilson Center. In 2008, this inventory listed hundreds of consumer products using nanomaterials, be they carbon nanotubes in construction products, titanium dioxide in cosmetics, or silver nanoparticles in food containers. But an inventory such as this one cannot be taken at face value. It relies on the self-declaration of companies that might decide to advertise a nanotechnology quality or, fearing potential health concerns, remain silent on the subject. The French cosmetic company L'Oréal provided a telling illustration of the strategic importance of the "nanotechnology" label. L'Oréal first used it for some of its products, before withdrawing it when considering it was more a liability than an asset.

The identification of the objects of nanotechnology requires that the problem of definition be solved, with or without connection with the problem of the potential risks of these objects. As the criteria defining "nano-ness" have not been determined, and the measuring instruments that could implement these criteria not standardized, one cannot rely on a stable technical infrastructure. This chapter describes attempts at defining nanomaterials in standardization and regulatory institutions that need to deal with poorly standardized measuring instruments, uncertain choices of criteria, and contradictory risk studies. The issue here is that of public intervention in a state of (ontological) uncertainty. This problem of definition is also a problem of political legitimacy. Why are the chosen definitions valued? By and for whom? For what objective? Is nano-ness to be related to potential risks? Defining nanomaterials is necessarily a task for which the ontological and the normative are brought together, or in STS analytical language, are coproduced (Jasanoff 2004). This calls for considering the construction of technical legal categories defining the existence of objects, not only as a classification work to analyze as such (Bowker and Star 1999), but also as a "constitutional" process, in the sense that it leads to a distribution of powers between decision makers, citizens and scientific experts, and of collective values expected to be pursued (Jasanoff 2011).

As in the other chapters of this book, I start the analysis by describing agencements, which are, in this case, the sociotechnical configurations that define nanomaterials, and connect these definitions to values to pursue and objectives to meet and to publics to whom they are addressed. Examining agencement here means analyzing the mode of existence of nanomaterials as a regulatory category, without supposing that there is a given reality that it should (and could) unproblematically describe. It is also a way of analyzing the collective and individual agencies that the categorization work produces, whether national delegations negotiating against each other in international arenas, stakeholders confronting one another in adversarial settings, or public and private actors jointly exploring the uncertainties of nanomaterials.

In the previous chapters, I have discussed sites where nanotechnology was problematized as an issue related to publics to inform and engage. Simultaneously, the sites analyzed in these chapters were also places where democracy was defined as a problem of citizen participation in public life. In the sites examined in this chapter, nanotechnology is made a regulatory problem, to be dealt with by the standardization or administrative tools of public institutions. This does not mean that while the previous sites were all devoted to "publics," the ones described here are all about the technicalities of expert work. On the contrary, they are different entry points to our more general objective: picturing the making of nanotechnology as a general entity coproduced with democratic order. These entry points directly connect democratic ordering with the issues related to the construction of markets, be they international, European, or national, within which new economic goods ("nano" products or "nano" materials) are expected to flow. Qualifying these goods as "nano" is a complex task, undertaken in various places. This chapter focuses on the work of definition of nanomaterials in different arenas—international, European, or national. The International Standardization Organization (ISO), as the central organization for international standardization, tackled the issue of defining nanomaterials. It did so by imposing a size restriction that made coping with technical uncertainties and international negotiation constraints easier. However, this size criterion did not allow U.S. and European regulatory bodies to address the potential risks of nanomaterials: the United States as well as Europe would rather assess them on a case-by-case basis. Still, the European case is a very specific one as new categories regulating definitions of nanomaterials are introduced. Finally, and as I will discuss at the end of this chapter, when public and private French actors introduced new categories, new definitions of nanomaterials were associated with experimental political practices.

International "Science-Based" Nanomaterials

Standardizing Nanotechnologies "by Science"

Crafting shared definitions of nanotechnology objects, and ensuring that they could circulate in global markets became an objective of international standardization. In 2007, a Technical Committee (TC) was created at the ISO to work on nanotechnology. ISO is an international meeting space for national standardization organizations, themselves hybrid institutions mixing private and public actors.[1] National delegations gather industrial representatives (usually via professional federations) as well as civil servants from the scientific policy and risk management sectors. The TC229 (Technical Committee 229) in charge of nanotechnology was created by officials in charge of national nanotechnology support programs, and the connection with the national science policy program remained strong. In 2010, for instance, the head of the U.S. delegation was Clayton Teague, director of the National Nanotechnology Coordination Office, the U.S. authority that coordinates all the federal programs devoted to nanotechnology.

At the May 2010 TC229 meeting in Maastricht, the Netherlands, I listened to a member of the Canadian delegation addressing a small group of scholars interested in standardization issues. He spoke about nanotechnology in those terms: "Nano is an abnormal group. We've never done this before. It's really about taking the beginning of the scientific basis to understand what we're talking about. (...) Usually, we're looking at products. But we're ignorant of what nanotech is."[2]

The ontological uncertainty of the field made it quite different from other industrial domains: industrial products supposed to be normalized were not easy to identify, and there was no such thing as a "nano" area with shared expectations and references. The TC229 first chose to set up a list of shared terms. Crafting definitions became a prerequisite to any of the other TC229 missions, especially the standardization of measuring instruments and methods for dealing with health, security, and environmental safety issues. Accordingly, the TC229 was separated into three working groups (WGs) performing specific tasks: defining chemicals (WG1), measuring them (WG2), and assessing risks (WG3).[3]

From Nanoscale to Nanomaterials

Within WG1, the definition of nanomaterials resulted from an iterative process, labeled "science-based" at ISO. Here, being science-based consisted in crafting basic definitions first, before elaborating more complex ones. Therefore, WG1 first defined "nanoscale," then "nano-objects," and finally "nanomaterials." In order to define "nanoscale," the 100 nm size limit, below which objects would be qualified as nano, was quickly accepted. It was already mentioned in national and international nanotechnologies reports, notably those of the U.S. National Nanotechnology Initiative (NNI),[4] of the British Royal Society[5], and of the OECD.[6] It also figured in TC229's mission statement: the committee was in charge of standardization "in the field of nanotechnologies, including a comprehensive and controlled matter on the nanoscale, typically but not exclusively, under 100 nm for one or several dimensions."[7] In these texts, the 100 nm size limit was considered "an order of magnitude," a "typical yet not exclusive" dimension. This size limit was used to define, in a synthetic way, a publicly funded research program while taking into account a range of scientific works converging in a particular direction ("new properties for dimensions in the range of 10 nm") rather than a boundary based only on laboratory instrumentation. That was the definition adopted by TC229's WG1.

Making the difference between what is nano and what is not consists in not being "too big." It is also about not being "too small." For if basic molecules enter the nanoscale, then what would happen to organic molecules, such as petrochemical products? They could become nano too, and nanotechnology would not be differentiated with other areas of the chemical industry. This is the reason why WG1 chose to set a 1 nm inferior limit to define the nanoscale. Still, some cases, such as fullerenes, proved problematic. These soccer ball-shaped carbon compounds (C_{60}) were synthesized and characterized by chemist Richard Smalley. For this, Smalley was awarded the Nobel Prize in Chemistry in 1996, and is now considered one of the founding fathers of nanotechnologies. A fullerene has a size smaller than 1 nm. Then, if the nanoscale is defined so as to exclude fullerenes, these compounds would not be considered nanomaterials. Yet, excluding fullerenes "would have been an aberration," as a member of WG1 put it during an interview.[8] Not only are they "the basic structure of all carbon-based nanomaterials" but also "the starting point of all nano programs." Being Smalley's major breakthrough, they have been used to demonstrate to science policy bodies the importance of nanotechnology.[9] Thus, it was unthinkable to exclude fullerenes from the nanoscale without ignoring a major part of nanotechnologies development programs. The nanoscale had to rely on science policy to be stable: any standardized category for nanotechnology would depend as much on development programs as on the small size of substances.

This prevented WG1 from defining the nanoscale limits too strictly. WG1 eventually settled on this: the nanoscale goes from "approximately" 1 to 100 nm.[10] Following the same logic they used for defining the nanoscale, WG1 then turned to "nano-objects," and defined them as substances having at least one dimension on the nanoscale. For instance, a nanoparticle was defined as an object with three dimensions on the nanoscale; and a nanotube as a tube with two dimensions (i.e., the transversal section) on the nanoscale (ibid.). The next step was nanomaterials, and was contentious.

The main issue was the following: should nanomaterials be defined as nano-objects, or should the definition of nanomaterials also include "nanostructured" materials (i.e., having nanometric structural regularities) as well as nano-objects? The discussion was about whether to extend the definition of nanomaterials. And this was a sensitive topic. For if defining nanomaterials meant targeting a class of chemicals for future regulation, then the extension of the category would imply increased constraints put on industries. A member of the French delegation whom I interviewed made the position in this debate a matter of economic interest. For him, the opposition of some national delegations—including the German one, where representatives of the chemical industry were numerous—to the extension of nanomaterials to nanostructured materials was nothing more than reluctance to paving the way to future regulatory constraint.[11] The narrative this delegate offered is consistent with the framework of the international negotiation at ISO, where each delegation is expected to defend the interests of its local economy. The ISO agencement made the discussion revolve around the negotiations between national delegations arguing for what was construed as their interests. This had consequences for the definition of nanomaterials that was chosen, which eventually made nanomaterials the addition of nano-objects and nanostructured materials.

A "Science-Based" Definition of Nanomaterials: The Impossible Property-Based Definition

For all the opposition to the extension of the nanomaterials category, stabilizing the definition of nanomaterials once and for all as a set of nano-objects and nanostructured materials was possible because it could be said to be "science-based." And this is a very important point. It implies, on the one hand, that the successive definitions crafted by the WG1 are logically connected to each other: if the "nano" nature depicts regularity on the nanoscale, then nanostructured materials have to be included. On the other hand, these definitions do not promote "political" purposes. The term "political" was used by members of TC229 to refer to regulatory decisions. As a matter of fact, the idea of imposing regulatory constraints on nanomaterials manufacturers was conceived by the international body as a national sovereignty initiative, with which international standards should not interfere. This implied that including nanostructured materials in the classification of nanomaterials could not be justified by the desire to regulate those materials. It was only the description of a "scientific"—and not "political"—criterion, namely nanometric regularities, be they manufactured or not, related to new properties or not.

Thus, the linear logic of the definitions of nanoscale, nano-objects, and nanomaterials made it impossible to define nanomaterials in ways that could be construed as "political." This was precisely the case of attempted definitions of nanomaterials based on properties related to size rather than size itself. For instance, researchers have provided a definition of inorganic nanoparticles "from an environmental, security and safety perspective" (Auffan et al. 2009). This particular approach brings technical development and risk assessment together without distinguishing those two aspects. This could lead to defining a "nano-object" according to "properties *related* to size rather than to size itself" (ibid., 641). These properties would be—among others—the reaction surface area, the ion release levels, the capacity of oxidation, whose impact on the toxicity of products can be measured.

Defining the fact of being nano according to properties was brought up during discussions in WG1 but nothing ever came of it. Properties related to size—including toxicological ones—vary from one chemical to another and from Company X's product to Company Y's product. Measuring instruments are not standardized and methods are heterogeneous to measure the criteria that could characterize properties related to size. The purpose of TC229's WG2 was precisely to work on measuring methods, but it could not contribute to alternative definitions, and not only for technical reasons. Indeed, if WG2 had selected a property along with a preferred measuring instrument in order to make it a basis for a definition, then the owner of the technology in question would have been favored at the expense of those who would have to buy it from that owner. This turned out to be quite an issue within international negotiations. Even more problematic was the fact that electing a criterion according to properties would have connected the problem of definition with the problem of risks—precisely what ISO considered a "political" move.

International Nanomaterials, International Negotiations

At ISO, the size criterion, a legacy of the public policy programs supporting nanotechnologies, became the only way of defining nanomaterials. By contrast, definitions that would have been based on size-related properties were unable to receive international standard status. The size criterion avoids examining the specific features of each material, whose properties depend on a great variety of physico-technical features. This is both a technical (the size of materials has to be measured one way or another) and a science policy criterion (public funding programs define a transversal sector, i.e., a sector within which materials measure "approximately" between 1 and 100 nm), which is not related to the elaboration of a restrictive regulation concerning nanomaterials. That is why the 1–100 nm limit could fit into the standardization body whereas alternative definitions, lacking the measuring infrastructure and threatening to mix the problem of definition work with a regulatory purpose deemed "political," could not. The size limit is indeed a "science-based" criterion, yet it is based on a science that has to do with science policy, international diplomacy, and technical instrumentation. International nanomaterials, as defined by ISO, are defined within an agencement characterized by the separation between a technical standardization-related expertise, disconnected from sovereign national choices possibly involved in the regulation of objects, and national positions defined by the economic interests they represent. The definition problem is addressed here as a "scientific" question, separated from all "political" considerations that could threaten the possibility of international consensus. This is a condition for the international organization to function as well as an outcome of its working processes, whereby national delegations competing for their national interests are careful not to favor one over the others.

A Case-by-Case Approach in the United States

Reconnecting the Problem of Nano-ness and the Problem of Risk

At ISO, the problem of the definition of nanomaterials was carefully separated from the problems related to the potential risks of nano substances. This had consequences: the chosen definition was merely conventional and of little help for public bodies attempting to locate substances for which the existing regulation could not take their risks into account. But as nanotechnology became an issue in public administration agencies, the problem of definition and the problem of risk evaluation could not be kept isolated from each other—and this made the issue of the identification of "nano substances" even more acute. Consider, for instance, the case of the U.S. Environmental Protection Agency (EPA). In 2004, EPA implemented a voluntary declaration program asking willing manufacturers to provide information about the physico-chemical properties of the "materials on the nanoscale" they used (EPA 2004). By targeting "materials at the nanoscale," EPA chose, as did ISO, to connect "nano-ness" with a size range. But the results were not satisfactory.[12] The question was, how to differentiate among substances with the same chemical composition but different sizes and shapes, and, accordingly, potential different profile hazards (for instance, carbon nanotubes)?

In 2006, EPA declared that the single nanoscale limit was not enough for defining new regulatory categories.[13] The agency considered that a substance only differing in size from a substance known to EPA would be considered as already "existing." This meant, for instance, that a titanium dioxide in the nano state was not different from its "non-nano" counterpart. The language of "existence" said there less a philosophical stance on the ontological nature of nanotechnology objects than a direct reference to the legal text framing the use of chemicals. Within the American legislation regulating chemicals, the Toxic Substances Control Act (TSCA), any substance listed on an inventory is regarded as an "existing" chemical whereas any substance that is not is considered a "new" chemical. Hence the question about nanomaterials: are they "new" or "existing"? Framed like this, the problem is that of a regulatory agency attempting to regulate hazardous objects. It is also that of private companies producing chemicals expected to be registered. Considering that no single size criterion could determine the novelty of chemicals, EPA answered these questions by looking at each separate substance that could be considered a "nanomaterial," and examining whether it could be said to be "new" or not.

This, more generally, was the approach undertaken by the federal agencies concerned with the regulation of nanotechnology. It was stimulated by the intervention of actors concerned about specific nanomaterials and attempting to make the regulation evolve. The case of silver nanoparticles is particularly interesting for that matter, since it offers an illustration of the processes through which the "nano-ness" of substances has been discussed in the United States.

Is Nano-Silver a New Object?

Silver nanoparticles—silver compounds made of about a hundred to a thousand atoms—were not the first substances mentioned in nanotechnology public policy programs. Contrary to carbon nanotubes, identified in the early 1990s, which quickly became major references in nanotechnology research and policy (e.g., Roco and Bainbridge 2003a, 2003b), silver nanoparticles were rarely mentioned in the early years of the construction of nanotechnology programs. The first concerns about nanoscale silver originated in the United States, when the Environmental Protection Agency was asked by the National Resource Defense Council (NRDC) to regulate a washing machine developed by the Korean firm Samsung because of its alleged use of so-called "silver nano," which was claimed to make the washing machine "anti-bacterial." The biocidal properties of silver ions were well known:[14] the "silver nano" would increase their degree. This situation led EPA to clarify its position regarding the "ion generating devices."[15] Samsung "silver nano" was indeed presented as a device that released silver ions at a regular pace. Worried about the "nano" label of the Samsung machine and the implications its initiatives about ion generating devices may have conveyed in terms of a potential step toward the regulation of nano substances, EPA made it clear that its objective was "not to regulate nanotechnology" (ibid).

The Samsung washing machine episode made nanoscale silver an object of public concern in the United States. The expression "nanosilver" then became widespread as a topic in public debate,[16] and started designating nanoscale silver compounds integrated into consumer products. Nigel Walker, director of the National Toxicology Program, said that the Samsung affair was the origin for the inscription of "nanoscale silver" in the nano-materials safety initiative in 2008.[17] The central question in this program—and the Samsung case is an illustration of it—was to assess the so-called "zero hypothesis": is nanosilver toxic because of the silver ions it releases (in which case its "nano-ness" does not transform the known biocidal effects of silver ions)? Or does it convey specific toxicological properties? In other terms, is the toxicity of nanosilver reducible to well-known toxicity of silver ions?

For our concern here, the zero hypothesis is particularly interesting, because it directly raises the question of the equivalence between nano and non-nano. It is an empirical entry point through which the potential nano identity of objects is discussed by regulators, environmental activists, and industrialists. When arguing about the validity of the zero hypothesis, these actors use operations that connect the identification of chemicals with legal initiatives and technical considerations. An example of such operations is provided by an initiative undertaken by the International Center of Technology Assessment (ICTA), a nonprofit organization that had been working on pesticides. Supported by a coalition of NGOs brought together by ICTA, a petition sent to EPA in 2007 asked the agency to regulate nanosilver as a pesticide.[18] For ICTA, the point was to prove that nanosilver was a new pesticide, not reducible to existing products using silver in a "non-nano" state.

In the American legislation, the text regulating pesticides is the Federal Insecticide, Fungicide and Rodenticide Act (FIFRA). Using pesticide law (i.e., FIFRA) rather than toxic law (i.e., TSCA) was a decision based on the difficulty ICTA perceived in the mobilization of TSCA in order to grant nanosilver a legal existence.[19] As noted earlier, entering nanosilver into the TSCA inventory of existing substances is not straightforward. Contrary to the case of fiber-shaped nanotubes, for which a physical criterion (e.g., atomic arrangement) can be used, nanosilver differs from silver ions only by the size of the set of silver atoms it is made of, which, as EPA made clear a year after the 2007 petition, is not sufficient basis to demonstrate the need for a new entry in the TSCA inventory.[20] In comparison, it was easier for ICTA to claim the novelty of nanosilver as a pesticide, since FIFRA deals with products and their properties, and not the chemical identities of substances. The distinction is interesting, since it points both to the ontological role of law, and to the specific problem of nanotechnology: is it a matter of size, or of properties related to size? Size was used, at ISO, as a technical descriptor unrelated to properties. Properties related to size, by contrast, connect nano-ness with the intended applications of substances. By choosing FIFRA, ICTA attempted to make size-related properties a basis for the regulation of new substances.

The argumentation ICTA used in the petition in order to demonstrate the novelty of nanosilver articulated legal and technical components. It referred to patenting practices as indicators of the novelty of silver nanoparticles. It mentioned scientific works that isolated silver nanoparticles. For instance, ICTA used a scientific publication that described the extraction of silver nanoparticles from the matrix in which they had been included.[21] The authors of this publication could then analyze the various shapes of the nanoparticles, and the effects they had on living cells. Through a study such as this, silver nanoparticles appeared as isolated substances, which differed from non-nano silver, and could even be differentiated from each other, based on their size or shape. Imaging technologies and physical tools of extraction were needed to perform such work. They complemented the legal tools that ICTA was using in the petition in order to isolate nanosilver as a new substance.

Isolated by ICTA in the petition, nanosilver could then be identified as an object NGOs needed to mobilize for since, as ICTA's argument went, EPA was reluctant to deal with it in spite of its specific identity. For ICTA, the mobilization of NGOs could make the regulation of nano substances move forward, that is, grant existence to previously not legally recognized nano substances. These actions implied alliances and collaboration among NGOs: ICTA initiated a partnership with U.S. and international consumer and environmental organizations and trade unions, which released "principles for the oversight of nanotechnology" in 2007.

Multiplied Existence of Nanosilver

In October 2009, I met Michael DiRienzo in his Washington, DC office. DiRienzo was the director of the Silver Institute, an organization representing and supported by companies of the silver industry. He was preparing the upcoming meeting of an expert committee, FIFRA Scientific Advisory Panel (SAP), commissioned by EPA in the wake of ICTA's petition. His comments on the petition were not moderate. He considered that the civil society organizations arguing for the regulation of nanosilver "hadn't done their homework." He was confident that the Silver Institute would convince EPA that nanosilver was nothing other than the well-known colloidal silver, that is, a solution of silver compounds of various sizes. He referred to the work done by the Silver Nanotechnology Working Group (SNWG), created by the Silver Institute, directed by a chemist based at the university of North Carolina, and comprised of approximately ten researchers. The communications of SNWG to SAP argued that "the majority of existing registered silver products [within FIFRA] are nano silver, including the algaecides and water filters that have been in use for decades. In fact all EPA registered silver products through to 1994 were nanoscale silver."[22] As colloidal silver had never been demonstrated to be hazardous, could be managed properly using the zero hypothesis, and was already regulated within FIFRA, SNWG stated that no new measures were necessary—a position that could be held by not drawing the distinctions ICTA did in the petition among shapes, sizes, and states of aggregation.[23]

The opposition between ICTA and the Silver Institute is easy to read in terms of the two groups' interests. In fact, each presented the other in these very terms, whether ICTA saw the Silver Institute as driven by the economic interests of companies wary of additional regulatory constraints (and the framed picture of Ronald Reagan on the wall of DiRienzo's office made it easy to adopt this narrative), or DiRienzo described ICTA's petition as an attempt to stir up irrational fears to attract funding and membership. These narratives are not for us to take for granted here. But they are outcomes of an agencement, typical of the American regulatory system, which results in private companies and civil society organizations arguing over the scientific validity of their positions within federal bodies of expertise.

The November 2009 meeting of the SAP was a place where such a debate took place. There, the problems of nano-ness were directly raised in association with the potential risks of the substance. Whether or not new risks require that new substances were created within the federal regulation was the initial question of the SAP:[24] was the toxicity of nanosilver reducible to well-known toxicity of silver ions? This initial question is exactly what ICTA and the Silver Institute disagreed on. But during the examinations presented during the SAP meeting, it was further complicated by the varieties of situations impacting potential risks. The panel members noticed that many of the products using silver nanoparticles comprised a wide distribution of particle sizes. And size was not the only parameter considered: the properties of the surfaces on which the silver compounds are deposited (for disinfection and sanitization applications) impacted the release rates of silver nanoparticles as the products were used; depending on where they circulate, silver nanoparticles may agglomerate, which could modify their toxicological properties.

The exchanges during the SAP meeting made nanosilver multiply: if differentiated according to their toxicological properties, nanosilver forms could be as many as the combinations of size range, agglomeration capabilities, and possibilities for bounding with nitrates or other natural substances in water. Consequently, the validity of the zero hypothesis (can one use silver ions data in order to evaluate the toxicological properties of nanosilver?), which was a key question in EPA commissioning SAP, was displaced. Not only did the available data appear insufficient for the panel members, but the relevance of the question itself appeared doubtful. Indeed, even if one considered that the mechanism determining the toxicity of nanosilver was the action of the silver ions it released, then the modalities of the circulation of nanosilver in the environment (or the human body)—which depend on physical and chemical characteristics, not necessarily the same across the range of nanosilver products—impacted the quantity and the frequency of the released silver ions. Thus, the opposition between "there are risks that are specifically linked to nanosilver" (ICTA's position) and "nanosilver risks are reducible to those of silver ions" (Silver Institute's position) could not hold anymore.

Faced with the proliferation of potential nano silver, and the impossibility of performing classical risk analysis, EPA did not attempt (as ICTA had) to make nanosilver exist as a new chemical substance within the FIFRA framework. The federal agency has continued after the SAP meeting to regulate companies' claims: if a company declares that a product has biocidal properties, it then must register it as a pesticide, whereas no specific requirement for nanosilver is specified. SAP suggested that future research should work on the physical and chemical characteristics of nanosilver (e.g., size, specific surface area, shape) and link them to the hazards of the substances (ibid., 37–38). Hence, the difficulty in dealing with the uncertain existence of nanosilver and the impossibility of controlling substances led EPA experts to call for "more science," more precisely "more predictive toxicology." The dynamics between the legal opposition among stakeholders and the (non) resolution of controversy by the call for science has been descried in other examples at EPA since the creation of the federal agency (Jasanoff 1990, 1992). In the case of nanosilver, it led EPA not to define the existence of nanosilver while simultaneously arguing for the necessity to do so, at a hypothetical future time when "more science" would be available. But nanosilver does not stand still in the products in which it is used, its fate in the environment and the human body is uncertain, administrative agencies (and industries buying it as raw materials) do not know where it is used, its existence is not defined, and companies' claims are a shaky ground for regulation making. Here, the call for "more science" has little chance to provide uncontroversial results.

The case of nanosilver provides an illustration of the quandary in which the case-by-case approach results. On the one hand it avoids introducing overarching criteria according to which one could decipher whether or not any substance is nano or not. But, on the other hand, it faces endless examination for each particular case, which may well result in the postponing of any regulatory choice. At this stage, it seems that there is a tension between two positions: either defining a general nano criteria inherited from science policy, and that separates the problem of the definition of "nano-ness" from the problem of risk regulation, or entering endless case-by-case examination. The former is what the international arena undertook, where it connected with the mechanism of international negotiations; the latter was adopted by the U.S. federal agencies, where it took the format of the stakeholders' negotiations.

Fighting over European Nanomaterials, Arguing about Europe

European Case-by-Case Approach

During a workshop called Safety for Success,[25] aiming to promote dialogue between the different stakeholders (member states, industries, associations) interested in the European regulation on nanomaterials, Cornelis Brekelmans, a civil servant of the Directorate-General (DG) Enterprise of the European Commission, explained in 2008 that "the regulatory framework covered the health risks" of nanomaterials. He quickly added that there was "a lot of work to be done to develop implementation rules in order to make the case by case approach work."[26] The "regulatory framework" mentioned by Brekelmans was based on the "Registration, Evaluation, Authorisation and Restriction of Chemicals" (REACH) regulation that frames the chemicals market in Europe. The major component of the REACH regulation is the "registration" process: for quantities over a ton per year, manufacturers have to submit a registration dossier to the European Chemicals Agency (ECHA) to prove they can master the risks. According to Cornelis Brekelmans, thanks to the "case by case approach," REACH could be applied to nanomaterials. His intervention announced the position the European Commission was about to make public in a Communication addressing nanomaterials. Published in 2009, this Communication stated that no new regulatory category was to be created for nanotechnology, and that the existing regulation had to be applied to each specific case. Proposing to adopt a case-by-case approach, the EC asserted that it was willing to seriously address the potential risks of nanomaterials, while not transforming the current regulatory system by creating a new size-based chemical category.

The European Commission's position directly resonates with the EPA's reluctance to constitute new regulatory categories for nanotechnology substances. For all the differences between the European and U.S. regulation of chemicals,[27] the case-by-case approach seems to be a shared method on both sides of the Atlantic. Yet there are important differences to consider. These differences pertain to the agencements that articulate the problematization of nanotechnology as an ontological enterprise about chemicals with the organization of public decision making. In the United States the case-by-case examination was directly connected with the functioning of the federal agencies, the tension between stakeholders of conflicting interests, and the role of science as a neutral resource to end conflict (and postpone regulatory choices). In Europe, the case-by-case examination was not conducted the same way, nor was it an uncontroversial approach within the European institutions.

Examining "Cases" through Regulatory Precaution

How does the European case-by-case examination work for nanomaterials? As is true of American regulation, companies need to know whether the substances they produce or use can be considered identical to previously registered ones. According to REACH, two substances are different if they have different chemical composition or if they show different physical properties among which are "elemental composition with spectral data, the crystalline structure as revealed by X-ray diffraction (XRD), [infrared] absorption peaks, swelling index, cation exchange capacity or other physical and chemical properties."[28] This list displays chemical and physical properties without discriminating among them. How should they be used in order to discriminate among substances? Clarifying its position about nanomaterials and REACH, the European Commission required that "all relevant information" be indicated in registration dossiers.[29] Yet what makes information "relevant," and whether new dossiers are needed, is precisely what is complex to decipher. Take, for instance, carbon nanotubes as produced in David Bertrand's French company. They may be rigid or not, single-walled or multiwalled, of various length and diameters—and the producer often does not know the exact profile of the substances he or she sells. The endless combination of properties, each of them potentially impacting the hazard profile of carbon nanotubes, might make regulatory categories endlessly multiply.

Examining these categories is precisely what the European Chemical Agency (ECHA), in charge of the implementation of REACH, has been undertaking. It has done so through collective projects gathering European and national experts from public bodies and private companies. Some of these "REACH implementation projects" were devoted to nanomaterials. Results released in 2011 showed that the case of carbon nanotubes could be dealt with by creating a "nanotube" category encompassing the variety of different tubes. But the situation was far more complicated for other nanomaterials, such as nanosilver or nano titanium dioxide, for which no consensus was reached on their identification with their bulk counterpart.[30]

Thus, the case by case consideration of nanomaterials within REACH made the European Commission extend the careful collective examination of each of the cases. It is, as such, revealing about the position the EC adopted on chemicals, and which places the precautionary principle at the core of its regulatory approach. This stance has been criticized on two sides. For some, the European approach to chemical regulation is a sign of undue public intervention in industrial regulation (e.g., Marchant and Mossman 2004). For others, "precaution" is employed as a catchword with little practical translation in terms of legal constraints exercised over private companies.[31] Yet neither of these positions accounts for the "regulatory precaution" through which the European institutions attempt to adopt a precautionary approach to potential health and safety risks by multiplying expert interventions, while at the same time using regulatory constraints with precaution, through multiple negotiations between public bodies and private actors (Boullier and Laurent 2015). This makes the European case-by-case approach different from the American one. Rather than a component of a regulation making process based on the negotiation between stakeholders and the call for science, the European case-by-case approach resulted in the proliferation of both the cases of different nanomaterials and the places where these cases had to be examined by European and member states' civil servants and representatives of private companies. This proliferation came at the price of the very possibility to settle the discussions by the introduction of regulatory constraints. The particularity of the European Commission's regulatory precaution is even more visible when one considers the opposition that this approach faced.

Case-by-Case vs. Nano Category: Regulatory Precaution vs. European Liberal Democracy

When it confronted the position of the European Commission on nano-technologies, the European Parliament opposed the case-by-case approach and declared that nanomaterials were to be regarded as a specific category.[32] Stating that "it did not agree with the Commission" on the fact that the existing regulatory instruments were sufficient to control nanotechnology substances, the Parliament attempted to create new "nano" regulatory categories. In 2009, it introduced a new amendment to the regulation concerning cosmetics. This amendment specifically targeted nanomaterials and imposed specific labels on cosmetics using nanotechnology. Here is the definition of nanomaterials it provided: "'nanomaterial' means an insoluble or biopersistant and intentionally manufactured material with one or more external dimensions, or an internal structure, on the scale from 1 to 100 nm."[33]

Through this initiative, the European Parliament introduced a definition of "nano-ness"—precisely what the EC and the American federal bodies had refused to do, and which ISO did by separating the problem of nano-ness from that of risks. But the Parliament did so in a very different manner from ISO. The definition it introduced only targeted "intentionally manufactured materials," that is, those that the regulation was expected to control. It used the 1–100 nm limitation along with the added terms "insoluble" and "biopersistant," thereby attempting to capture potentially hazardous materials. Hence the difference with the "science-based" ISO definition. For the European Parliament, the problem of definition was also a problem of risk regulation. This required making nano-ness visible for the European consumer and for the European regulator.

The approaches of both the European Commission and the European Parliament have been criticized. While the EC's regulatory precaution made the number of different cases multiply with little constraining power over industrial practices, the Parliament's position introduced a definition that is questionable. For legal writing constraints,[34] the modulating adverb "approximately" (used in the ISO definition to qualify the size limits) was not used in the cosmetic regulation. Consequently, European NGOs have worried that manufacturers wishing to elude the regulation could use substances bigger than 100 nm (110 nm, for instance) that would nonetheless have enhanced reactivity because of their size. Accordingly, the European Environmental Bureau (EEB) argued for a 300 nm upper size limit.[35]

But our interest here is not to evaluate which of the two approaches (if either) is the most efficient. It is far more interesting to notice the close articulation between the ontological task undertaken within the European institutions and the modalities of the European public intervention in technology regulation. Through the opposition between the European Commission and the European Parliament, two ways of conceiving the sources of the European democratic legitimacy appear. While the European Commission's regulatory precaution made the collective exploration of "cases" through collective discussions the basis of its intervention on nanomaterials, the European Parliament adopted a liberal perspective, within which the elected representatives of the European citizens ensured that Europeans could become informed consumers, choosing whether or not they wanted to buy and use specifically labeled "nano" products. The opposition between the EC and the Parliament is about the political nature of the European space, and the sources of its democratic legitimacy. It is also about the economic nature of this space, whether a new market is created where products labeled as "nano" and informed consumers can meet, or multiple markets for each different "nano" substance are constituted.

An Ambiguous Attempt at Defining Nano-ness

Following its initial attempt through the cosmetic regulation, the European Parliament introduced more complex definitions in the food and biocide regulations, again through amendments added to the text prepared by the European Commission and the Council. These amendments enforced labeling and introduced definition criteria that were more complex than those found in the cosmetics regulation. Thus, they required that any material exceeding 100 nm yet having "characteristic properties of the nanoscale" would be considered a nanomaterial. Among those properties, the amendment mentioned "those related to the large specific surface area of the materials considered; and/or specific physico-chemical properties which are different from those of the non-nanoform of the same material."[36]

This evolution is significant. It displays some of the attempts undertaken to grasp nano-ness through new properties related to size rather than size itself. It further develops an intermediary position, between the conventional construct with no physical meaning ("1–100 nm") and the scientifically more rigorous and never-ending examining of specific cases. Within the European institutions, this intermediary position was directly related to the position of the European Parliament, opposed to the EC's regulatory precaution and concerned with the information provided to the European consumer. At the Parliament's demand, it also resulted in a "recommendation," published by the European Commission, and related to the definition of nanomaterials. Released in 2011, the core of the definition was the following: "'Nanomaterial' means a natural, incidental or manufactured material containing particles, in an unbound state or as an aggregate or as an agglomerate and where, for 50 percent or more of the particles in the number size distribution, one or more external dimensions is in the size range 1 nm–100 nm."[37]

The number size distribution was the topic of heated debates between industries and NGOs in determining a threshold including more or fewer substances in the definition. When I met people working at the European Environmental Bureau (EEB), a federation of European environmental organizations lobbying in Brussels, the negotiations about the recommendation were well advanced, but the opposition not solved. While the EEB referred to a report written by the Scientific Committee on Newly Identified Health Risks (SCENIHR) to argue for a 0.15 percent threshold,[38] the industrial federation pushed for a much higher one.

The eventual decision was seen as a compromise by the EEB and private companies, which satisfied neither of the two sides. Also a compromise was the relation between the problem of nano-ness and the problem of the risks of nanotechnology objects. Materials covered by the definition were said to be "not more hazardous as such than larger but otherwise identical materials."[39] When the EC further explained that "whether a nanomaterial is hazardous will only be determined as part of a risk assessment" (ibid.), it separated the problem of definition from the problem of risk evaluation—precisely what the Parliament sought to bring together. Yet the definition introduced criteria such as the specific surface area that are directly related to the reactivity of nanotechnology substances. More than that, it explicitly mentioned the possibility of modifying these criteria in case of "concerns for the environment, health, safety or competitiveness."[40] Contrary to other affirmations in the very same recommendation, these moves articulated the problem of existence with the problem of risks.

How to make sense of this EC definition of "nanomaterials"? First, one needs to understand that the recommendation was eventually of little consequence, since it is not related to any legal obligation. It did not change the position of the European Commission regarding REACH (i.e., the case-by-case approach). The recommended definition could thus be seen as a gesture on the part of the EC, which little implication in terms of regulatory choices. Yet it also complements the description of the controversy about the regulatory existence of nanomaterials within the European institutions. The opposition between the European Commission and the European Parliament concerns whether and how nano-ness should be identified. It also concerns the sources of the European democratic legitimacy. The EC's recommendation on nanomaterials can be seen as an attempt at developing a middle-of-the-road position, agreeing with the Parliament on the definition of nanomaterials, and maintaining ambiguity about the articulation between the problem of existence and the problem of risks. It is also a sign of the uncertain nature of the scientific and political resources expected to contribute to the construction of the European regulatory space.

Experimenting with French Regulatory Categories

Governing Uncertainty

In October 2009, I traveled back to Paris from Brussels with Arila Pochet, an official at the French ministry of health. We were returning from one of the Safety for Success meetings that the European Commission organized from 2007 to 2011 to discuss the risks and regulations of nanomaterials. Pochet had been involved in the European REACH negotiations about nanomaterials and was a member of the French delegation to ISO TC229. I had interviewed her before, and we continued the discussion on the Thalys high-speed train. She told me about a project she was about to launch at the French national standardization organization, AFNOR, a mixed public-private entity in charge of industrial standards. Her project would develop a "nano-responsible standard," defining principles that should be followed by manufacturers who want to produce, use, or sell "responsible" nanomaterials. Her idea was to help companies deal with the regulatory uncertainty about the status of nanomaterials. She wanted to offer them the possibility of acting in precautionary ways even though nanomaterials were neither properly identified nor defined. She meant the project to be inclusive, and called on private companies, government bodies, and civil society organizations to participate. She asked me to join as an external expert—an offer I accepted as an (explicit) opportunity for participant observation.

The standard gradually took shape during a series of meetings at AFNOR, and adopted the form of a list of questions addressing every single step (design, production, transformation, use, and disposal) in the life of a product labeled as "nano" when companies claim to be leveraging properties related to the size of components. These questions were related to nanotechnology substances characterization, techniques of use, and reprocessing, as illustrated below by examples from an internal working document:

• What are the determining physico-chemical characteristics of the produced substance? Size of substances? Number size distribution? Shape? Surface reaction area?
• Could the production process release nanoparticles into the atmosphere? What are the materials concerned? Have risk studies already been conducted?
• Can we assess the risk of witnessing unexpected releases, dispersals, or exposures during the cycle of life of a product?

Manufacturers would then have to reflect on the future of their products, the point here being to encourage standard users to reflect on elements they might not have considered and to offer them ways of managing uncertainty (e.g., confinement, informing customers, switching to a better known substance).

The nano-responsible norm can appear as a way of *not* defining nanoness but dealing with the ontological uncertainty of the domain. It was thought of as a flexible tool, expected to evolve as new technical issues or new social concerns would emerge. When I participated in the AFNOR meetings devoted to the nano-responsible standard, I witnessed lengthy discussions about how to craft these questions, and how to associate them with a working standard, possibly leading to a certification. Companies were reluctant to envision potential constraints, and were worried about the relevance of the questions mentioned in the standard. Meanwhile, representatives of civil society organizations were concerned about the possibility of transforming this initiative into a platform for the collective examination of safety concerns.

These discussions were taken to another level when France proposed the project to the European Committee for Standardization (CEN). In doing so, the objective was to widen the scale of the initiative undertaken in France. It was successful, as AFNOR, leading the project and making it a central component of its approach, was elected to lead the secretariat of the CEN Technical Committee in charge of nanotechnology in December 2010. The project then entered a long phase of collective negotiation, in which I did not take part, and which in 2014 was still inconclusive, when I talked to Pochet four years after our discussion on the Brussels-Paris train. She was still excited about a project that had become European, but skeptical about its eventual outcome. She told me that risk–benefit analysis had become the accepted framing for discussion, whereas the initial French initiative sought explicitly alternatives to risk–benefit in order to cope with situations where neither risks nor benefits could be easily evaluated. Within CEN, which is in charge of operationalizing some of the policy directions defined by the European Commission (Borraz 2007), the French initiative had to be reformulated so that it could become an acceptable topic of negotiation among the other European delegations.

The final outcome of the process is less important here than what it says of the treatment of nanotechnology objects in France. The nano-responsible initiative was an attempt at making the uncertainties of nanotechnology governable. It relied on new and imperfect instruments, which traveled uneasily outside of France and could be subject to criticisms from within the plurality of actors it wished to include. As such, it is a component of an approach that manifests itself in other initiatives in France, some of which aim to establish regulatory categories for nanomaterials.

A Regulatory Category for Nanotechnology Objects

The national environmental consultation process that led to the organization of the national nanotechnology debate (see chapter 3) had another consequence. In the piece of legislation it led to, the French government was mandated to introduce a declaration of "substances in a nanoparticulate state" (*substances à l'état nanoparticulaire*). In this 2009 law, this expression was not clarified, and for people involved in the ministries in charge, it aimed to target materials that would release nanoparticles in the environment.

In February 2012, the government released the decree implementing the law, and further detailed it in a subsequent regulation (*arrêté*). After previous and mildly successful attempts at asking companies to declare their nanotechnology-related activities on a voluntary basis, France then became the first country to introduce a mandatory declaration of nanotechnology substances. In doing so, the French public administration had to define the substances it asked companies to declare. In this definition, and as other regulatory bodies attempted, the objective was to connect the problem of existence (what does make a substance "nano"?) with the problem of risk (how to identify potentially hazardous substances?). The definition took inspiration from the European recommendation, and used the same size distribution criterion, and the same specific surface area criterion. But it differed in important ways, since it only targeted manufactured materials that were either nanoparticles or expected to release nanoscale substances. The initiative of the French administration articulates the definition of a new entity with a legal innovation. The regulatory category was constituted in the French law with an unprecedented amount of technical details. A public agency called ANSES, overseeing health and safety, was in charge of controlling the mandatory declaration—a new task for an agency with little enforcement power.

The French initiative was met with skepticism by scientists, industrialists, and legal scholars (Lacour 2012). Consider, for instance, the reaction of the nanotechnology commission of AFNOR, the French national standardization organization. Commenting on preliminary versions of the decree, the AFNOR nanotechnology commission sent a letter to the French government in which it remarked that "substances in a nanoparticulate state" was not a known expression at the international level. It stated that "no standard method whether published or under study at the ISO, at the CEN (the European Committee for Standardization) or at AFNOR, could be suggested as reference method to support the implementation of this decree."[41] These words are important. They made explicit one of the crucial difficulties of any attempt at writing sophisticated definitions of nanomaterials. In the French definition (as in the European recommendation), criteria related to size distribution and to the surface of substances were used. Yet no standardized measuring method existed for these criteria. The critique, emanating from a standardization body especially aware of the issue, was not incidental. It meant that the regulatory category, for all its sophistication, could not rely on a stable network of standardized instruments and measures.

But difficulties did not end there. How, for instance, were the French public bodies in charge of the control of the mandatory declaration expected to enforce it? Few resources available and unclear legal consequences were only one part of the issue. Consider, for instance, a company willing to declare its use of "substances in a nanoparticulate state." If it buys raw materials from suppliers outside of France, then the company might well ignore the technical details required by the mandatory declaration, or even whether or not the products it buys fall under the definition or not.

Extending the Domain of Public Expertise to Impure Categories

Considering these difficulties, it may be tempting to disregard the mandatory declaration initiative, or to be critical of it. Yet within our exploration of the problematization of nanotechnology as a lens for the analysis of contemporary democracies, one needs to identify what this initiative says about processes of democratic ordering. The challenge is the same as in the case of the national nanotechnology debate. In this latter case, I did not attempt to evaluate whether or not this initiative was "participatory enough," but I described the problematization of nanotechnology as an issue related to publics, and the associate problematization of the French national public as an additional topic of the centralized public expertise. One can adopt a similar perspective with the substances *à l'état nanoparticulaire*—a public intervention to which the conclusions of the national debate made explicit reference.

The technical and social difficulties of the French mandatory declaration were not ignored by the civil servants in charge. They knew that the measuring infrastructure was not there, and that they had little power over foreign firms distributing materials on the French territory. Yet they concluded that what was required, in this situation, was a public intervention for the state to know at least in a rough manner what was going on in the nanotechnology domain. When I met the head of the *bureau des substances chimiques* ("chemicals office") at the ministry of ecology right after the publication of the decree, he was explicit on these points: "I am perfectly aware that the category is not pure, we don't have everything we would need. ... But we knew that when we decided to go on with the substances in a nanoparticulate state. We know it's an impure category, that we will not be able to control everything and everywhere. But it is a way of starting to identify what is going on, it is a necessary means for the public administration to grasp these objects."[42]

That the "substances in a nanoparticulate state" was an "impure category" within the French regulation is telling. The expression points to the attempt of the French centralized administration to make uncertain objects governable, indeed to make ontological uncertainty itself governable. In intervening to define new entities with an impure category, the French public administration was also adopting a different approach than the European Parliament. Rather than defining in order to label consumer products for consumers to choose to buy them or not, the *substance* initiative sought to gather information for the sole sake of the centralized expertise. This is the role of ANSES, the French public health agency, in charge of gathering and analyzing the collected declarations. Ultimately, the mandatory declaration is thought of as an instrument of knowledge for the government to use to identify (if in an incomplete manner) the landscape of nanotechnology industrial activities.

This attempt is tentative, as yet to be stabilized within an agencement that could distribute roles and responsibilities among public and private actors, and define regulatory categories in uncontroversial ways. Whether or not the regulatory innovation will further develop the ability of the French public administration to mobilize technical expertise is not settled yet. It depends on whether or not the administration will require companies to play the game of mandatory declaration, and will be able to convince the French public that it is able to identify and control nano-technology objects requesting regulation. But there is an additional public the French public administration has been attempting to convince, at the European level. As soon as nanomaterials were discussed within REACH, France figured among the countries that argued for the creation of a new "nano" category in the European regulation. Accordingly, the *substance* mandatory declaration was conceived as a demonstration in front of Euro-pean witnesses of the possibility for such a mandatory declaration to be introduced (ibid.). This further qualifies what was at stake with the regula-tory innovation: as it was attempting to extend the realm of its interven-tion to uncertainly defined objects, the French public administration tried to convince internal and external publics of the validity of its interven-tion. Thus, one can consider this legal innovation as a disruptive experi-ment, attempting to redraw the modalities of public intervention in technological issues while maintaining the ability of the state to act on behalf of the general interest. Just as the French state attempted to include an expertise on publics with CNDP (see chapter 3), it also attempted to incorporate undefined objects into the domain of the governable. The problem of definition, here, is also a problem of the internal and external legitimacy of the French administrative power. In chapter 3, I described the national public debate on nanotechnology as a state experiment. The nanotechnology substances initiative shows that this state experiment has another component, related to the construction of impure regulatory categories.

Defining Nanomaterials, Problematizing Nanotechnology

In June 2011, Andrew Maynard, an American toxicologist who had been involved in numerous studies about the risks of nanomaterials, published a comment in *Nature*. Entitled "Don't Define Nanomaterials," this piece opened with the following statement: "Five years ago, I was a proponent of a regulatory definition of engineered nanomaterials. I have changed my mind. With policymakers looking for clear definitions on which to build 'nano-regulations,' there is a growing danger of science being pushed aside" (Maynard 2011, 31).

Maynard was commenting on the recommendation of the European Commission about the definition of nanomaterials. He remarked that it was the result of a "policy decision," which, for him, was antithetical to what should be a "scientific" one. After having spent years of studying nanomaterials and arguing for their regulation, Maynard was concerned about criteria such as the 100 nm size limit or the 60 m2/cm3 surface limit, which he considered were at best outcomes of collective negotiations, at worst arbitrary technocratic decisions. But his opinion piece was pessimistic. For he did not see the case-by-case approach undertaken in REACH and within the U.S. regulatory framework as satisfactory either. He called for "adaptive regulations" that could make new properties related to size rather than a mere size criterion the basis of regulatory categories. This would be, for him, the real "science-based" approach.

Maynard's comment about the definition of "nanomaterials" illustrates the quandary public administrations find themselves in, concerning the risks of nano-objects. While it is necessary to make these objects visible for public bodies to regulate them, the general "nano" category is controversial. How to define objects whose only similarity lies in scientific policy programs? How to implement definition criteria when there are hardly any infrastructures in place to compare physico-chemical parameters? The agencements constructed to answer these questions articulate definition criteria, policy objectives, public concerns to address, and expectations of acceptable behaviors on the part of concerned actors. They all rely on expert advice, and all claim to be "science-based" and "policy relevant." But what counts as "scientific," *pace* Andrew Maynard, and what makes these initiatives "policy relevant" vary greatly across the sites examined in this chapter, according to the connections they draw (or refuse to draw) between the problem of definition of "nano-ness" and the problem of the risks of nanotechnology objects.

The main issue at ISO is to provide working international expertise intended to help create a global market for nano while being careful not to tread upon national prerogatives. ISO focuses on the characterization of nano-objects, regardless of their future uses and the issues they might create. It separates the problem of definition from the problem of risks, which makes it impossible to craft definitions based on size-related properties rather than size itself (precisely what Maynard called for). In the international standardization organization, expertise is supposed to be produced in a space freed from regulatory considerations and from the potential futures of nanotechnologies. By contrast, national and European regulators connect the problem of definition with the problem of risks. In Europe and the United States, the way of doing so (at least initially) is to examine each separate substance. This distinguishes two issues: the examination of risks inherent to objects on the one hand, to be dealt with on a case-by-case basis, and the construction of nanotechnology as a global program worthy of a new category on the other hand, promoting these very objects for the sake of future developments. This asymmetry between the public treatment of risks and benefits has been described in other technological domains, most notably biotechnology (e.g., Jasanoff 2005). Here, it is integrated in different problematizations of nanotechnology. While the U.S. regulatory system makes the examination of different cases a matter of legal procedures, the European discussion about nanomaterials soon turned into a problem related to both the potential definition of nano-ness and the nature of the European political and economic space. How to act in uncertain situations is also an issue for the French administrative actors. But they deal with it in different ways. Engaging in regulatory innovation, they attempt to reinvent the modalities of the intervention of the centralized state, so that uncertainty becomes a manageable domain. This requires that impure categories are crafted—"impure" at both technical and social levels.

In the sites examined in this chapter, nanotechnology is made a problem of government—concerning uncertain entities expected to circulate in international, European, or national markets. In these sites, defining "nano" substances means acting on and through markets thanks to various types of regulatory interventions. None of these sites proposes a definitive solution to the regulation of nanotechnology objects. Indeed, they all face pervasive difficulties, whether they introduce displacements in the modalities of public decision making and expert interventions, or reproduce existing regulatory practices. It is precisely by analyzing these difficulties as moments of problematization that one can locate sites where the rules governing standardization, regulatory, and expert institutions are questioned, restated, or displaced. As such, the implications of these rules for national sovereignty, national or European citizenship, and the legitimacy of collective decision making clearly appear.

5 Making Responsible Futures

The Futures of Nanotechnology

Nanotechnology programs were defined as plans for the future from their inception. As the American nanotechnology policy reports, particularly those concerned with the "convergence" of nanotechnology with biotechnology and information science, discussed the future developments of nanotechnology in connection with other technological advances, they used emphatic affirmations such as: "We stand at the threshold of a new renaissance in science and technology, based on a comprehensive understanding of the structure and behavior of matter from the nanoscale up the most complex system yet discovered, the human brain" (Roco and Bainbridge 2003a, 1).

Promised was a world where "people may possess entirely new capabilities for relations with each other, with machines, and with the institutions of civilization" (Roco and Bainbridge 2003a, 22). Announcing a "new renaissance" questions the functioning of democracy: what roles are citizens expected to play in the making of this bright future? How determinist is the vision of the future presented? What are the possibilities for public action to shape the future of nanotechnology?

The actors involved in the definition of nanotechnology's programs likewise raise these questions. While the "new renaissance" discourse triggered them in the first place, they also brought up concerns regarding the more mundane components of nanotechnology, such as applications in medicine and construction, where safety issues could arise. Overall, the proponents of nanotechnology considered it crucial for the success of the nanotechnology programs that public support was ensured, and public controversies anticipated. Contrary to previous experiences, such as happened with GMOs, nanotechnology would require, so the argument goes, a "responsible" approach, within which risks would be taken into account and publics invited to participate. Ultimately, the "responsible development" of nanotechnology would ensure the smooth construction of markets based on technological development.

Responsible development was heralded as a central principle for innovation in nanotechnology. As defined by the director of the U.S. National Nanotechnology Initiative (NNI), responsible development refers to all the operations undertaken to mitigate the potential risks of nanotechnology and maximize their benefits, while informing the public about both risks and benefits.[1] In Europe, and as this chapter will show, nanotechnology appeared to be an experiment in the definition of a more general objective of "responsible research and innovation."

The responsible development of nanotechnology forces us to question the production of "responsibility," both at an individual and collective level. The notion of agencement is useful for that matter, as a way of not predefining what "responsible" means, and of identifying the particular agency of individuals and collective entities (such as private companies or public administrations) responsibility entails. Rather than using predefined criteria according to which one could judge whether or not nanotechnology development is indeed "responsible," describing the agencements that makes objects, people, and organizations responsible offers an entry point in the analysis of sites where nanotechnology is problematized as an issue of future making.

Accordingly, this chapter considers successively American and European initiatives meant to make nanotechnology developments responsible. These initiatives originate from science policy institutions wishing to anticipate potential concerns, while ensuring the development of nanotechnology in acceptable ways. As such, they directly raise questions related to the expertise needed to conduct anticipatory interventions.

An American Expertise for the Making of Responsible Futures

Liberal and Conservative Ethics Struggle with Nanotechnology

Talking about a "new renaissance" in the definition of science policy programs is not neutral. As formulated in American converging technologies programs, it directly echoed the central themes of a school of thought known as "transhumanism," which contends that humans need to use technological progress to "enhance" themselves. Scholars inspired by transhumanism have had strong interest in nanotechnology, which resulted in the American *Converging Technologies for Improving Human Performances* report.[2] Some of the transhumanist thinkers are concerned about the need for "appropriate information" in order for each individual to decide whether or not she would want to be "enhanced" (Bostrom 2003). Others argue for a "democratic transhumanism," which would ensure that every type of being, be they enhanced or disabled, straight or cross-gendered, human or animal, could live according to his or her personal choices (Hughes 2004). But in all cases, at the heart of transhumanist thinking is a reflection about technological development and the modalities of social intervention in it.

This reflects the more general issue of the "implications" of nanotechnology, which adopts several versions across the American science policy scene. At one end of the spectrum, it points to the need for society to adapt to a technological development that is deemed to be unstoppable. At the other end, it relates to the collective construction of technological programs. The variety of these positions—not always acknowledged, and sometimes leading to contradictory statements in the science policy literature—resulted in long discussions in science policy arenas, which the existing expertise for managing the implications of technological development was not sufficient to deal with (see Fisher and Mahajan 2006a). A major component of this existing expertise was ethics, which had reflected upon other technological domains. The life sciences, for that matter, are a particular telling area. Bioethics has been used as a way of reflecting on technological development and its implications on value choices, and has become an expertise regularly mobilized for the management of medical practice or the reflection on technological advances impacting the very nature of life (Evans 2000, 2002). This can take a liberal or a conservative format. In the former, decisions are delegated to the individual, expected to make autonomous decisions, and ethics is there to ensure that the conditions of individual choice are met—possibly including conditions of justice in a given community.[3] In the latter, overarching values such as "human dignity" determines the acceptability of technological development (President's Council on Bioethics 2008). Whether it takes the "liberal" or "conservative" format, bioethics is based on a dichotomy between scientific facts and value according to which they are evaluated. In both cases, bioethics is based on an expertise about "values," whether it uses, in the liberal version, general principles such as autonomy, benevolence, nonmaleficence, and justice (these four principles define what has been known as "principalism"), or, in the conservative version, concepts such as "human dignity."

Nanotechnology was another area of intervention for both conservative and liberal ethics. But both sides struggled with it. They were asked to reflect on concerns that were linked to future developments promised, often in emphatic ways, in science policy programs and had trouble locating the area where their expertise could be exercised. Thus, the President's Council on Bioethics—created by President George W. Bush in 2001 and headed until 2005 by Leon Kass, a known critic of biotechnological innovation and proponent of human dignity[4]—struggled to locate the "specific ethical issues" of nanotechnology, although it had taken a vocal stance against the use of technology "beyond therapy" (President's Council on Bioethics 2003). At the liberal side of the spectrum, ethicists who had been working on other technological domains with the tools of principalism created a *Nanoethics Group* and crafted an approach based on a permanent "catching up" of ethics as nanotechnology would develop (Lin 2007; Moor 2005; Moor and Weckert 2004)—an approach that by definition is bound to wait for technical apparatus to materialize before developing any ethical argumentation. For the people involved in ethics expertise, nanotechnology appeared as a challenge, and a difficult one. They were to leave the safety issues to technical examination of risks, and wait for ethical issues to arise as nanotechnology developed (and produced, for instance, applications in the field of human enhancement).

Considering these difficulties, it is not surprising that alternative approaches were proposed. During a public hearing organized by the President's Council on Bioethics (PCB) in 2007 about the ethical issues of nanotechnology, UNESCO director of the Division of Ethics of Science and Technology Henk ten Have insisted on the need to introduce mechanisms for dialogue among scientists, humanists, and citizens. For ten Have, ethics was tightly involved with other social actors in the construction of institutions that could exercise an "ethical vigilance" on emerging technologies. This, however, was not considered part of PCB's role, as one of the members of the council stated: "Now, many of the other problems you mentioned, Dr. ten Have, seem to me to be extremely important, but I view them more as issues in politics or issues in general prudence, things that should be done, for instance, to re-insert science in the political community, for instance, to regenerate trust. But, again, I'm not ... maybe I'm blind to this. I don't see the specific ethical issue that would require reflection."[5]

Ten Have's suggestion indeed led to not waiting for facts to be solidified enough in order to mobilize values; he proposed, rather, that ethics should intervene in the collective construction of nanotechnology itself (which then comprised science policy programs as well as material objects). This differed from the human dignity approach of PCB. That Ten Have provided no concrete example of infrastructure able to enact such a "constructionist ethics" did not help convince the members of the council to abandon the human dignity approach.

A New Ethics for an Emerging Technology?

This failed intervention in front of the PCB was not the only attempt at displacing the usual ethics expertise in the American science policy scene. Philosopher George Khushf, who contributed to reports published by the National Academy of Science, argued for a "situated ethical reflection" (see Khushf 2004, 2007a–c) that would contribute to the development of nano-technology objects alongside scientific research works.[6] Khushf's propositions were included in reports published by the National Academy of Science and the National Research Council.[7] But they remained only a minority component of a much more diverse nanotechnology policy literature that, under the general theme of the responsible development of nano-technology, allowed Khushf's and the nanoethicists' arguments to be simultaneously acceptable.[8]

Some isolated experiences along Khushf's lines of thought were conducted after these initial developments. Thus, when I talked to Khushf during an interview, he mentioned a study undertaken by Christopher Kelty, an American anthropologist, which he considered a good illustration of what his approach could be.[9] Kelty had been interested in the work of the Center for Biological and Environmental Nanotechnology (CBEN) at Rice University (see Kelty 2009). CBEN does research on nanomaterials and their environmental applications, such as water treatment. One of the researchers at CBEN is Vicki Colvin, whose work on fullerenes is widely recognized, and who played a major role in the definition of the federal nanotechnology policy. Testifying before Congress, Colvin advocated the inclusion of the "impacts" of nanotechnology in the federal programs so that the risks of nanotechnology might not become risks *for* nanotechnology.[10] Rather than analyzing the risks of fullerenes once their potential use had been defined, Colvin's approach consisted in characterizing the toxicity of the substance as a function of its structure. As other properties of fullerenes were linked to their nanometric scale, their toxicity might play interesting roles, for example, in destroying tumor cells. Like the other properties of nano substances, toxicity needed to be controlled. In CBEN research projects, toxicity thus became, according to Kelty, a fundamental property of the material, as much as its surface area, its atomic mass, or its density. As such, it was yet another parameter on which to play in order to design nano substances with interesting functions. The "implications" of nanotechnology were then integrated within the very practices of scientific research, and within the material itself. The approach was labeled "safety by design" by Vicki Colvin, as the design of the material comprises its toxicological properties.

"Safety by design" echoes the attempts to propose property-based definitions of nano substances described in chapter 4. In this perspective, there is no difference any more in the work about "safety" and the work about "ethics." Safety by design deals with the construction of nano substances and products. Here, ethics is not separated from scientific practice. Health risks are ethical issues, in so far as they imply the identification of the substances' properties, meaning, their "characterization," a term used in the physical sciences and by George Khushf to label his perspective on the ethics of nanotechnology. The issue at stake here is no longer to separate "principles of action" (e.g., bioethics principles, or human dignity) from the content of the action, but to open up the construction of all the aspects of scientific practice and technical developments, be they decisions occurring in the laboratory, characteristics of technical systems, modes of collaborations among disciplines and actors, expected usefulness, or future distributions of applications. One can locate this alternate approach elsewhere, in social scientific experiments undertaken by scholars with ambition to develop a new methodology for technology assessment that would be adapted to nanotechnology, and to which I now turn.

Making Experimental Nanotechnology Futures

Political scientist David Guston and Daniel Sarewitz proposed in the early 2000s to develop "real-time technology assessment" (RTTA), which would aim to "integrate social science and policy research natural science and engineering investigation from the outset" (Guston and Sarewitz 2002, 2). Guston and Sarewitz pursued the STS analysis of the coproduction of science and society, and took inspiration from European methodologies for technology assessment, such as Constructive or Participatory Technology Assessment (CTA and PTA) (Schot and Rip 1997). Like CTA, RTTA was meant to integrate technology assessment into the making of technologies. Like PTA, it hoped to involve stakeholders and publics in the reflexive and deliberative construction of technology. Guston and Sarewitz were careful to differentiate their approach from existing methodologies. They argued that RTTA was about the production of new knowledge rather than the experimentation about new technologies, that it would develop tools for the analysis of the evolution of public values and concerns, and that it sought to integrate retrospective case studies with prospective explorations (Guston and Sarewitz 2002, 6). Guston and Sarewitz used nanotechnology as a domain where RTTA could (and should) be implemented. RTTA was further institutionalized when the Center for Nanotechnology in Society (CNS) was created at Arizona State University, under a National Science Foundation grant established after the 2003 21st Nanotechnology Research and Development Act. Directed by Guston and hosted by the Consortium for Science Policy and Outcomes directed by Sarewitz, CNS received the biggest award granted by NSF for social science research in nanotechnology, and became by far the main project within the NNI in the social and ethical implications part of the program. CNS-ASU is not the only NNI-funded project expected to ensure that the "implications" of nanotechnology are adequately dealt with. But it is the main component of a set of initiatives through which, according to their promoters and to the director of NNI himself, "nanotechnology is becoming a model for addressing the societal implications and governance issues of emerging technologies generally" (Roco et al. 2011, 406).[11]

Guston and Sarewitz presented RTTA as a step forward after previous attempts to link social science and scientific research, and thereby ensure that scientific and technological development was conducted in responsible ways. One of these previous attempts was the "Ethical, Legal and Social Implications" (ELSI) program of the Human Genome Project (HGP), which, according to Guston and Sarewitz "had not been well-integrated into either the science process or the R&D process" (Guston and Sarewitz 2002, 3). It is worth discussing the ELSI program of the HGP, since it sheds light on the objectives of RTTA as operationalized in the Center for Nanotechnology in Society.

The Human Genome Project allocated 3 percent of its funding to the study of "Ethical, Legal and Social Implications" of genetic research. The ELSI program, famously backed by DNA discoverer and Nobel Prize winner Jim Watson at the launch of the HGP project (Jasanoff 2005, 177–180), led to the examination of ethical issues connected to human genome. Projects were funded to study the "ethical implications" of human genome research, the organization of such research, and the construction of science policy. As its first director Eric Juengst put it, the ELSI program was meant to address "the virtuous genome scientist's professional ethical question: 'What should I know in order to conduct my (otherwise valuable) work in a socially responsible way?'" (Juengst 1996, 68).

For all the enthusiasm of its initiators, the ELSI program was heavily criticized. A source of tension was the conflicting demands it was submitted to. The ELSI program was expected to ensure its objectivity (i.e., that it was not captured by political interest). Juengst, a bioethicist directly involved in the expansion of principalism as a tool for the objectivity of ethics advice (Evans 2002, 24, 162), was concerned with the production of independent knowledge. Juengst insisted on the quality and intellectual independence of the ELSI research, as he responded to critics questioning how "objective" ELSI grantees could be about any issue that bears on genome research, when their funding is provided by the genome research community on the assumption that genome research is a good to be protected (Juengst 1996, 70).

But the problem of objectivity was deeper than that of the source of funding because the ELSI program was also asked to be "politically relevant," as a report from Congress stated (U.S. Congress House of Representatives 2012). This meant that it was supposed to provide advice that could be directly translated into policymaking. These competing expectations resulted in a complex institutional history, during which the program faced multiple changes of status, in order for the institutional body not to be absorbed by alleged political interests.[12] Critics of ELSI were the basis for the creation of the National Bioethics Advisory Commission (McCain 2002, 132; U.S. Congress House of Representatives 1992), which institutionalized the principles of bioethics, as instruments for the advisory committee (Evans 2002). This evolution was much to the dissatisfaction of Eric Juengst, for whom the role of ELSI was to generate knowledge and a community of specialists able to use it, with no formalized process of connection between the production of objective knowledge and that of policymaking (Juengst 1991, 1994, 1996).

The dynamics at play here is remarkably similar to that of the expertise of federal bodies for scientific and technical issues. As Sheila Jasanoff has argued, the production of scientific advice in the U.S. administrative circles has had to deal with concerns about the objectivity and neutrality of expertise: the federal agencies that were the most explicit in separating the "policy" role from the "scientist" ones were those that faced destabilizations and accusations of producing an expertise that was overpoliticized (see Jasanoff 1990, 1992).[13] Such a tension was clearly at play in the case of the former Office of Technology Assessment (OTA), expected at the time of its inception in the 1970s to provide both "independent and "policy-relevant" advice. This eventually caused its elimination in a later period marked by severe cuts in the federal budget, as OTA proved unable to demonstrate the link between its expertise and law making, precisely because of the institutional construction of its neutrality (Bimber 1996).

As it appears through these episodes, the American expertise on the societal implications of science and technology is traditionally based on two dualisms. Not only are social norms and moral values to be separated from scientific facts in order to mobilize an ethical expertise independent of the question being examined, but the ethical expertise also needs to be separated from decision-making processes. Bioethics functions on both separations (whether under its "liberal" or "conservative" versions). Proponents of RTTA have been challenging the first dichotomy. In basing their reflections on a critique of HGP's ELSI program, they attempted to rethink the second.

The main critique that RTTA scholars addressed to the HGP ELSI program was indeed that it had "no policy relevance." For instance, Daniel Sarewitz and Ira Bennett, one of his colleagues from CNS, wrote about the failure of the ELSI program to "link ELSI research to policy decision processes."[14] The "no policy relevance" argument is debatable. At the very least, it would be vigorously opposed by Eric Juengst (Juengst 1996). Yet albeit its (probable) simplification, it was mobilized by CNS scholars as a useful counterexample to make their objectives explicit. Rather than developing research projects that would analyze the societal implications of nanotechnology for the sake of it, they would develop a technology assessment that could be fed into nanotechnology policymaking. Thereby, they would pave the way for a new Office of Technology Assessment, which could avoid the fate of the first OTA thanks to the combination of "policy-relevance" and "quality research."

The way to do so, for the proponents of RTTA, was to refuse the fact/value dichotomy and argue that it would intervene in the very making of nanotechnology objects, concerns, publics, and futures, and thereby ensure the "relevance" of their approach. But they also needed to demonstrate the quality of RTTA research, and its ability to provide expertise for the making of science policy. This could be done by making CNS a (social) scientific demonstration. In a small-scale environment, researchers would experiment with RTTA, and eventually demonstrate that a new office of technology assessment based on RTTA would be viable. The space of the demonstration was then an isolated one, which could be, in some instances, an actual scientific laboratory, and, in others, a totally different locus. In any case, it was supposed to contribute to policymaking by demonstrating the value of RTTA on the scale of the social scientific experiment.

Several instruments were put in place in order to do so. I already described one of them in chapter 3. The National Citizens' Technology Forum, a multisite citizen conference organized in 2008 on the topic of converging technologies and human enhancement, was conceived as a demonstration of the value of the consensus conference format for the engagement of the American public in discussions about future technologies and as a social scientific instrument through which deliberation dynamics could be studied. I will comment on two other initiatives meant to operationalize RTTA: the integration of social scientists in nanotechnology laboratories, and the making of scenarios about the potential developments of nanotechnology.[15]

Embedding Human and Social Scientists in the Laboratory

Some of the CNS researchers have been involved in a project to "embed" humanists and social scientists in a scientific laboratory. The project was based on the experience of a researcher, Erik Fisher, who was an "embedded humanist" at the Thermal and Nanotechnology Laboratory of the University of Colorado between 2003 and 2006 (Fisher 2007).[16] There, Fisher participated in various laboratory projects, talking with scientists and asking questions about their practices. He was interested among others in a project consisting of carbon nanotube synthesis in silica tubes ("tubes in tubes"). Applications for this project were being explored at that time; people mentioned for instance industrial applications for heat transfer. Following this project, Fisher had repeated discussions with the person in charge of the available technical options. For instance, as the project leader was about to use the usual catalyst, Fisher asked whether another one would be possible. The discussion that followed led them to consider the possibility of an iron nanoparticle solution, which eventually would be both more efficient for the synthesis of nanotubes, and less risky in terms of its toxicological impacts. Thus, Fisher argued that the embedded humanist experience contributed to rendering visible for the scientists themselves the microdecisions that are taken during the mundane course of research, and that the scientists' activity might be made "reflexive," in the sense that the everyday practices of scientific activity could be denaturalized thanks to the presence of the humanist, and potentially open to interrogation and reconfiguration. Eventually, Fisher expected the intervention of the humanist to transform the very outcomes of scientific process—a result of the "embeddedness" that the silica tubes story was meant to be a demonstration of. The "embedded humanist" thus hoped to perform a "midstream modulation" of nanotechnology research.[17] Its "midstream" quality was defined as such: "Viewed this way, the midstream corresponds to the implementation stage of a large, distributed, and dynamic decision process. For simplicity, upstream decisions may be characterized as determining what research to authorize, midstream decisions as determining how to implement R&D agendas, and downstream decisions as determining whether to adopt developed technologies" (Fisher, Mahajam, and Mitcham 2006, 490–491).

Through the metaphor of the "stream," the embedded humanism initiative could be inscribed in the whole RTTA project alongside "upstream public engagement" (conducted through NCTF) and downstream "societal implications research" (ibid., 493).

The work of the "embedded humanist" is meant to render nanotechnology problematic—in other words, as an entity of individual reflection for the scientist, and collective discussion with the humanist. It also implies a transformation of roles: that of the humanist as well as that of the scientist. The former does not hold "values" or "principles" which she could mobilize to study the "implications" of scientific research. Her own anthropological description contributes to the scientific project. The latter is led to denaturalize and question her everyday practices.[18]

In 2009, Fisher received funding from NSF for a project devoted to "socio-technical integration research" (STIR). STIR took the notion of embedded humanism to another scale. It coordinated about a dozen graduate students "embedded" in nanotechnology laboratories in ten different countries, who were asked to contribute to the project with "narratives of embeddedness," in which the "modulation" of scientific research could be made explicit, in the guise of Fisher's early experiments.[19] The number of embedded humanists was higher than Fisher's initial attempts, but the logic was the same: the objective was to describe and act on actual nanotechnology research practices in a selected number of laboratories, and ultimately to demonstrate the value of midstream modulation by gathering empirical cases where embedded human and social scientists transform research outcomes.[20]

Scenario Writing

Another instrument used at CNS has been scenario writing. Scenarios were conceived by the leaders of the scenario project at CNS as the basis of a work of collective reflection that aimed to explore what the future of nanotechnology could be, and make it a topic of pluralist deliberation. As Cynthia Selin, a member of CNS and leader of scenario projects put it:

The question that immediately arises from this mandate is: how to study and encourage deliberation of implications of something that has yet to occur? That is, nanotechnology is largely about potential and future deliverables, promising to be revolutionary. But given the inchoate form of it, there are no completely reliable and grounded ways to talk about implications. This situation poses challenges for the social scientists who have been summoned to go into the lab, talk to policy makers and engage the public about nanotechnology. They must confront the future. (Selin 2009)

Scenarios were conceived as an answer to these challenges. CNS members wrote the initial scenarios. They chose to focus on themes that had been discussed in the NNI works on the societal implications of nanotechnology, and that were present in the scientific, as well as in science fiction and popular literature.[21] The scenarios comprised the following examples:

• "Living with a brain chip": a brain chip delivers information inside the brain during the sleep of the user.
• "Automated sewer surveillance": a sequence technology is used to analyze DNA fragments in used waters, thereby permitting a control of populations.
• "Disease detector": a device measures the protein rates and detects abnormal levels even before the appearance of illness symptoms.

When I visited CNS-ASU for a few months in 2007, the project was just starting, and initial scenarios had been written and illustrated. NSF reviewed the activity of the center during my stay at ASU: one of the issues the evaluation raised was the plausibility of the scenarios. As Selin explained to me at that time, the NSF reviewer "wanted to know that (the scenarios) did not come out of nowhere."[22] Accordingly, she devoted much time and energy to solidify a process that could ensure the "plausibility" of the scenarios, and, therefore, the validity of the method she was developing. Scenarios, for her, were useful tools. She still had to demonstrate both their quality and their usefulness for the participants in the projects she was leading, and for funders interested in the policy-relevance of RTTA.

These demonstrations were conducted through the inclusion of review processes, whereby scientists commented on the scenarios in process. Scenarios would then be posted on the Internet for online commenting and discussed by the panel members of the National Citizens' Technology Forum on converging technologies (see chapter 3). They were sent to scientists, industry groups and NGOs representatives, who were asked to participate in online discussions.[23]

In these projects, the scenario was conceived as a way to not accept the dichotomy between "reality" and "science fiction" in order to make issues related to nanotechnology development explicit. For instance, participants in collective discussions would interrogate the types of market and social relationships an illness-tracking device might construct, should it become widespread. But for CNS members, scenarios were not only tools meant to stimulate a collective identification of the societal implications of nanotechnology. They were also expected to intervene on futures. They had the potential to reorient attention and modify action, as scientists and other participants reflected on the potential development of nano products. Reorientation and modification could only occur for the limited numbers of participants in the scenario projects organized at CNS. But RTTA scholars could then use these attempts in their research work, where, by interviewing participants and accounting for the gradual construction of scenarios, they could demonstrate the value of scenario making for the exploration of nanotechnology's societal implications and the modulation of participants' opinions and practices. In that sense, the mobilization of scenarios did not follow the approach undertaken by the nanoethicists, based on the correct representations of nanotechnology's facts. Nor did it operationalize a collective construction of nanotechnology programs. Here, the scenario was a basis for a collective reflection in the isolated setting of the social scientific laboratory, which could then demonstrate to academic colleagues and policymakers the usefulness of scenario writing for the making of nanotechnology's responsible futures.

Expertise for U.S. Responsible Futures of Nanotechnology

The objectivity of the expertise on the societal implications of technology is an important issue in the U.S. discussions about the making of responsible futures of nanotechnology, and a long-term concern of American science policy. To the concern about the objectivity of ethical advice is added the question of its "policy relevance," which renders problematic the link between the production of ethical or social scientific knowledge and policymaking. Nanoethicists have tried to replicate the liberal bioethics style of argumentation on nanotechnology. Both nanoethicists and the conservative ethicists at the President's Council for Bioethics were unsuccessful in intervening in nanotechnology. Based on values to be mobilized once scientific facts are established, ethicists were bound to wait for the nanotechnology objects to materialize.

As it refuses the separation between risk and ethical issues, and questions the link between (social) scientific expertise and public decision making,

CNS's real-time technology assessment shifts the discussion away from the opposition between liberal and conservative ethics. It attempts to get rid of the distinction between "ethics" or "social science" and "science," between "principles" and "research projects." In order to do so, it separates the small-scale experiments expected to demonstrate the value of this new approach to the actual making of science policy program. Going from the small-scale and demarcated experiment to the visible results expected to enlighten policymaking requires that the promoters of the experiment perform demonstrations. Displaying the outcomes of CNS's experiment to funders is an important aspect of RTTA. Its proponents organize meetings in front of U.S. Congress nanotechnology caucus, are attentive to presenting the results of their work to policymakers, and display their experimental results in numerous professional nanotechnology conferences.

In the sites examined so far in this chapter, nanotechnology is problematized as an issue of future making, while the nature of the expertise to mobilize in order to make technological development acceptable is simultaneously discussed. The construction of a legitimate expert intervention for the definition of collective values is a permanent problem to deal with, which is framed in the language of expert knowledge and objectivity. Objectivity thereby appears as a central concern, whether it is based on the principles of bioethics, which nanotechnology rendered more complex to apply, or on the laboratory setting that is RTTA. As such, the discussions about how to govern nanotechnology's future development in a responsible way serve as empirical entry points for a reexamining of a central ingredient in the construction of legitimate public intervention within the American democracy, namely the objectivity of expert decisions, and their ability to demonstrate that they are not captured by special interests. Turning to the European case, the next section will illustrate another approach, in which the concern for the responsible futures of nanotechnology is integrated in the construction of nanotechnology policy programs.

European Values for Responsible Nanotechnology Futures

While the United States was crafting a federal initiative for public support of nanotechnology research, the European institutions were launching initiatives in order not to be left behind (so the argument went). The 2004 Communication of the European Commission that presented the "European strategy for nanotechnology" made it clear that "one of the crucial differences between the EU and our main competitors is that the landscape of European R&D in nanotechnology risks becoming relatively fragmented with a disparate range of rapidly evolving programmes and funding sources" (European Commission 2004, 8).

In order to deal with fragmentation,[24] "integration" was heralded as a key concept. The notion of integration, in European policy language, has both a political and a moral undertone. It opposes fragmentation in that it points to the need to bring together the European member states' national initiatives within the European research space, in order to make the EU a "knowledge-based economy," as defined by the Lisbon strategy of 2000, and to use nanotechnology in order to shift "from a resource-intensive to a knowledge-intensive industry." But integration also means that common values are to be included at the core of nanotechnology policy.

This meant that particular areas of nanotechnology research were to be pushed forward, such as applications of nanotechnology for environmental purposes (e.g., nanomaterials for water treatment, nanomaterials for construction allowing energy savings) and health purposes (e.g., nanovector for drug delivery, nanoparticles as tracking devices inside the body).[25] The "integration" objective, however, meant more than targeting application sectors for nanotechnology research. It also involved the mobilization of science policy instruments expected to coordinate the work of the different member states, and making operational the "European values" and "European principles"[26] that were expected to be at the heart of the Lisbon strategy. This section describes several of the science policy instruments through which these European values and principles are supposed to bear on nanotechnology. Thereby, it analyzes the ways in which European actors attempt to make responsible nanotechnology futures. Although the concerns for the social and ethical issues of nanotechnology circulated from the United States to Europe, they were dealt with differently within the European institutions than in the United States. As I will discuss, nanotechnology appears as an experiment in the making of the European research policy, and, more generally, in the making of Europe itself as an integrated political and moral entity.

Responding to the American Nanotechnology Programs

As the American proponents of nanotechnology celebrated converging technologies and its applications for personal enhancement in a report entitled *Converging Technologies for Improving Human Performances* (Roco and Bainbridge 2003a), concerns for the responsible development of nanotechnology attracted interest in Europe, and led the European actors to problematize the future in different ways from their American counterparts. Thus, the 2004 *Converging Technologies—Shaping the Future of European Societies* report, to which philosopher Alfred Nordmann was a rapporteur and which was commissioned by the Directorate-General for Research and Innovation of the European Commission, was conceived as "a European response to the American NBIC (Nano-Bio-Info-Cogno convergence) program" (Nordmann 2004). Nordmann had been one of the early initiators of research projects in the ethics of nanotechnology in the United States.[27] When he moved back to Europe in 2004, he participated in the development of an approach summarized as "CTEKS," that is "Converging Technologies for European Knowledge Societies," which was supposed to characterize the European take on the development of nanotechnology. Nordmann recalled the influence of a European civil servant at the Directorate-General for Research and Innovation in the following terms:

One of the first readers of the NBIC report was Mike Rogers, at the time program officer at the Foresight Unit of the Directorate General Research of the European Commission. (...) Rogers suggests two ways in which the Commission ought to go beyond the U.S. report. The first way is to place a greater emphasis on the social sciences and humanities and take a more comprehensive approach to the cognitive sciences. The second way *is to integrate this convergence within European values to allow for the acceptance of the emerging technologies.* (Nordmann 2009, 287; emphasis added)

Being a former secretary of the European Group on Ethics (EGE) in science and new technology, Mike Rogers had been accustomed to working with "European values." More than that, he framed the whole process of defining a European approach to converging technologies and nanotechnology as an attempt to integrate "European values" in responsible science policymaking.

Hence, the perspective outlined in the CTEKS report proposed to develop converging technologies for "European values" such as solidarity, sustainable development, and the mutual production of social expectations and technological development. This made CTEKS an "integrated approach" and nanotechnology an exemplary case for the development of European democracy. As nanotechnology's objects were still to be crafted, and science policy programs to be defined, so the development of nanotechnology was an opportunity to "invite and empower" "European societies." Accordingly, the CTEKS report offered a skeptical account of the American version of converging technologies. In opposition to its perceived libertarian tone, the CTEKS report proposed no less than "a new social contract between science and society" intended to answer a series of "challenges," such as the "development of local solutions that foster natural and cultural diversity," the "promotion of sustainable development, environmental awareness, precautionary approaches," and the "empowerment (of) citizens and consumers to understand, use and control CTs and to maintain a sense of ownership" (Nordmann 2004, 54). This European take on converging technologies argued for the mutual construction of science and society, both as an empirical reality and as an objective for the making of European science policy. The report suggested the use of a series of instruments, including "European design specifications for converging technologies," an international "code of good conduct," the mobilization of the European Group on Ethics, and "foresight tools" that included "vision assessment" and "deliberations about the visions that underpin the development of CT" (ibid., recommendation 9).

Rather than an obligatory mobilization of society in order to realize the potential of autonomous technologies targeted to human enhancement (which were seen as characteristic of the American approach to converging technologies), the CTEKS report called the European nanotechnology policy a "collective experiment" within the shared development of converging technologies. The CTEKS report problematized nanotechnology as a matter of constructing a common European future. Here, the problem of the European future was defined in the terms of a common responsibility to establish a collective European entity based on shared values. Thereby, the report linked the European nanotechnology policy to the more general approach to EU research policy, that of the Lisbon strategy, and that of the European Research Area, which was expected to be "deeply rooted in European society" and which "should experiment with new ways of involving society at large in the definition, implementation and evaluation of research agendas and of promoting responsible scientific and technological progress, within a framework of common basic ethical principles and on the basis of agreed practices that can inspire the rest of the world" (European Commission 2007, 10).

Hence, the early formulations of the European nanotechnology policy introduced values and principles that would have to be integrated in the making of nanotechnology policy. The 2005 *Action Plan* that defined the main directions of the European nanotechnology policy followed up on this approach. It echoed some of the recommendations of the CTEKS report in calling the European Group of Ethics to "carry out an ethical analysis of nanomedicine," which was expected to "identify the primary ethical concerns and enable future ethical reviews of proposed N&N" (European Commission 2005, 9). The *Action Plan* suggested "embodying common shared principles for R&D in nanotechnology in a voluntary framework (for example, a 'code of good conduct')." (European Commission 2004, art 5). The later Framework Programmes, which defined the European research policy, proposed to operationalize these principles in defining goals such as "the environment as a whole: energy efficiency and sustainable energy production and the emergence of sustainable products," and "applications in the health-care field and development of nano-analytical tools."[28]

In defining a European approach to nanotechnology, the CTEKS report and the subsequent science policy documents made "integration" a central objective. What was to be integrated was the set of European values that were to be brought to bear on technological development, as well as the diversity of member states' research policies on nanotechnology. Making the future of nanotechnology a problem of integration connects political and moral aspects, and pertains, more generally, to the nature of Europe as a consistent political and moral space. How are the European institutions expected to meet the integration objective? Following the CTEKS report and the nanotechnology policy initiatives, several instruments of the European research policy were mobilized to do so.

European Principles for a European Expertise in Ethics

Following a request present in the nanotechnology *Action Plan*, the European Group on Ethics (EGE), an advisory body to the European Commission, published an *Opinion* on nanomedicine in 2007. In this text, "pluralism" was considered as one of the "characteristics of the European society" that was "mirroring the richness of its traditions and adding the need for mutual respect and tolerance. Respect for different philosophical, moral or legal approaches and for diverse cultures is implicit in the ethical dimension of building a democratic Europe. This is relevant also for the moral controversies prompted by nanomedicine" (European Group on Ethics 2007, 44).

For all its apparent political correctness, such unspecific language has important consequences. The "respect ... for diverse cultures" implies that the construction of a "democratic Europe" cannot be based on the solidification of rules that would contradict member states' "philosophical, moral or legal approaches." This is a variation on the subsidiarity principle, according to which the scope of interventions of the European institutions is limited to objectives that cannot be reached by member states alone. Within the subsidiarity principle, matters of ethics and value are delegated to member states, and the European institutions cannot impose constraining actions.

How is the EGE expected to contribute to the integration objective then? The EGE has been subjected to two kinds of criticisms, which directly oppose one another. On the one hand, scholars have blamed it for not being able to constrain technological development, thereby letting the European Commission impose for unique guiding principle the competitiveness of the European economy for the sake of market development (Tallacchini 2006; 2009). But for others, the EGE is going too far in influencing the European research policy,[29] as in the case of embryonic stem cells, in which the European Commission, following the advice of the EGE, restrained its funding of human embryo research produced in Europe (Plomer 2008; see Salter and Jones 2002). For the former critics, the EGE cannot define any other common European value than that of market development. For the latter critics, it introduces unduly constraints disregarding the subsidiarity principle. Both groups of critics point to a European mode of government that uses moral values as vehicle for political and moral integration. They differ on their evaluation of this mode of government, but rather than evaluating it from the outside, it is more interesting for our concern here to characterize the "European approach to ethics" that appears through the work of the EGE. The expression was used in a 2000 report in which the group described its role and duties (European Group on Ethics 2000). The 2000 report, written when the EGE was compelled to make its positions clear regarding European funding for stem cell research, listed principles that should be applied in European ethics, among them "human dignity," "individual freedom," "principle of solidarity," "freedom of research," "safety," "responsibility" and "transparency" (ibid., 11). That this list is long and heterogeneous is significant. The European approach to ethics does not stabilize a set of principles, nor does it predetermine the ways in which they should be applied. Rather, guided by a constant and dual concern for subsidiarity and integration, it redefines their contents for each new case.

The EGE's 2000 report introduced principles and values that were not predetermined and left open the boundary between what is defined at the European level and what is left to member states. It was a step in the making of a politics of common European morality. The 2007 nanomedicine *Opinion* was a case where this politics was exercised. The *Opinion* restated many of the values and principles mentioned in the 2000 report, asked for the safety of nano products to be ensured, and, pursuing CTEKS' critical perspective of the American approach to the future of converging technologies, appeared skeptical of applications of nanotechnology for "human enhancement." It stated that projects meant to enter this domain were not to be given priority over others, and that, at any rate, the distinction between "medical and nonmedical use" should be made clearer by European research in ethics.[30] Hence, EGE pointed to domains of technological activity that it considered outside the scope of the European values. But the *Opinion* required other science policy instruments to further implement the perspective it defended. One needs to connect the EGE with other instruments meant to provide European institutions with ways of influencing member states, research institutions, and individual scientists to act in accordance with European values.

A "Code of Conduct for Responsible Nanosciences and Nanotechnologies Research"

Making member states, research organizations, and scientists responsible was the objective of a Code of Conduct (CoC) that the European Commission released in 2008, after an initial proposition in the 2005 nanotechnology *Action Plan*. Directorate-General for Research and Innovation officials regularly presented the CoC as the most visible attempt to define a "European approach to ethics," which would "promote dialogue" while "not imposing some forms of ethics rather than others" (von Schomberg 2009). One recognizes here the language of "pluralism" that, in the European context, pertains to the issue of subsidiarity.

A consultation procedure and discussion resulted in the list of the "core European principles" that individual scientists, research institutions, and member states abiding by the CoC were expected to follow. The CoC mentioned seven of these "principles": "meaning," "sustainability," "precaution," "inclusiveness," "excellence," "innovation," and "accountability." Among the "prohibition, restrictions or limitations" the CoC introduced were

4.1.15 N&N (Nanoscience & Nanotechnology) research funding bodies should not fund research in areas which could involve the violation of fundamental rights or fundamental ethical principles, at either the research or development stages (e.g. artificial viruses with pathogenic potentials).

4.1.16 N&N research organisations should not undertake research aiming for non-therapeutic enhancement of human beings leading to addiction or solely for the illicit enhancement of the performance of the human body.

4.1.17 As long as risk assessment studies on long-term safety is not available, research involving deliberate intrusion of nano-objects into the human body, their inclusion in food (especially in food for babies), feed, toys, cosmetics and other products that may lead to exposure to humans and the environment, should be avoided. (European Commission 2009, 9)

The CoC was a vehicle for the nonconstraining integration of European values into the practice of scientific research. Thus, a laboratory abiding the CoC principles would not only apply the principle of "inclusiveness" and "accountability" by communicating about its research, it would also refuse to undertake "deliberate intrusion of nano-objects into the human body," and would not develop brain implants or nanodevices for drug delivery if meant to propose "nontherapeutic human enhancement (...) leading to addiction or solely for the illicit enhancement of the performance of the human body."

The CoC could not rely on a stable infrastructure that could have defined "nano-objects" in an unambiguous manner, and the boundaries between "licit" and "illicit" enhancement were not clarified. During a public workshop devoted to the code and organized in May 2008 by the DG Research and Innovation, I heard many criticisms about the unspecific language of the CoC and skepticism about the practical significance of the prohibitions it advanced. Physicist Richard Jones, then an advisor for nanotechnology for the British Engineering and Physical Sciences Research Council, later questioned the practical significance of pointing to the "accountability" of the individual researcher toward future applications of his or her work (Jones 2009). But what is interesting for our concern here is not to evaluate the efficiency of the Code, but to analyze it as a site of problematization. The code made nanotechnology an issue about values as well as a problem of European political and moral integration. This problem was expected to be dealt with in a flexible manner. Thus, the code delegated to project coordinators and scientists the reflection on both the appropriate domains of research in nanotechnology, and the practical details of research practice, without introducing mandatory actions or constraining requirements. Member states were recommended by the European institutions to "be guided by the general principles and guidelines for actions to be taken, set out in the Code of Conduct for Responsible Nanosciences and Nanotechnologies Research (...), as they formulate, adopt and implement their strategies for developing sustainable nanosciences and nanotechnologies (...) research" (European Commission 2009, 3). The CoC and the texts through which the European institutions recommend its use are parts of a European agencement within which integration functions through the nonconstraining diffusion of common values, themselves loosely defined, and outside of the regulatory area. Another component of this agencement is the ethics review process, which projects applying for European funding are required to submit to.

Integrating Ethics in Research Projects

Projects requiring European funding have to pass an "ethics review," which has direct consequences for the conduct of research. For instance, the ban on human embryonic stem cell research in European research projects (as recommended by the EGE) was enforced through the ethics review process. Ethics reviews are performed by expert panels, and coordinated by officials at the "governance and ethics" unit at the Directorate-General for Research and Innovation. They are based on the legal European requirements (e.g., directive on data protection, clinical trials, or animal rights) and national legislations.

The concerns nanotechnology raised needed special attention on the part of the experts reviewing the projects. In a context of uncertainty about both the potential risks and ethical concerns of nanotechnology, and about the national initiatives undertaken to deal with them, the ethics review process would introduce flexible references to a precautionary approach. It invited researchers to use the Code of Conduct, and attempted to ensure that minimal precautions were undertaken. This excerpt from an interview with D.K., an official in charge of the ethics review process at the European Commission is telling:

DK: If there is another concern, the experts are telling us. Because of the uncertainty surrounding, for instance, the use of nanomaterials. What is the effect on the scientists themselves? Is there stricter safety regulation needed, at the level of the lab? There, we rely on the safety regulation that exists in member states.

BL: But in the case of nanomaterials, there is no specific regulation?

DK: Right, to the best of my knowledge, there is none. So they refer to the general rules for safety and, hem, common sense. They refer to the precautionary approach. They suggest appropriate safety procedures, like the limitation of exposure.[31]

The case of nanomaterials illustrates how the principle of subsidiarity is operationalized at the level of the ethical review process. When no European or national legislation exists, but experts nonetheless identify a concern regarding European principles (e.g., "safety" or "transparency"), then the ethical review process leads researchers to adopt concrete actions (in the example considered here: "safety procedures" and "limitation of exposure").

Thus, the ethics review is a vehicle for the integration of the European space of scientific research, and an instrument of the European politics of morality. The integration is moral in that it relates to the harmonization of values deemed European. As opposed to technological domains in which member states have adopted nonharmonized positions and where research projects might abide by different ethical guidelines according to those of the member states they choose to follow,[32] nanotechnology offers an opportunity to ensure a minimal consistency across the moral space of European research. Integration is also political in that it characterizes an articulation of harmonization and subsidiarity, which attempts to govern technological development without introducing constraining legal provisions.

Ethics review processes are ways of integrating ethics within research projects, and, more generally, of integrating the moral and political space of the European research. But some aspects of nanotechnology proved more complex to advise the project leaders on. In some cases, experts from the ethics review panels would point to the potential implications of nanotechnology, or research leaders would raise issues about the potential implications of their work, when dealing, for instance, with brain implants offering potential control of the individual, or safety risks impossible to evaluate in a case of uncertainty about the characteristics of nano substances (see chapter 4).

Sometimes at the advice of the ethics and governance unit, "ethics boards" were constituted within research projects. For example, Nano2Life, the European network devoted to nanobiotechnology research (see chapter 1), comprised an ethics board, which was composed of eight members who included scientists, philosophers, social scientists, and religious leaders. Among the objectives of the board were the "evaluation of general and prospective ethical and social questions raised by nanobiotech Research & Development projects," the "monitoring of projects initiated by the network," the information of researchers and students about ethical issues, and the "dialogue with the public to identify ethical concerns of the European citizens." For instance, nano brain implants were examined within the Nano2Life network as part of an evaluation of "human enhancement technology." But rather than an ethical argumentation based on a set of stable principles independent of the issue on which they are applied, the ethics proposed by Nano2Life pushed for the integration of ethics in the conduct of a project. Such an integration had various implications. First, it implied the continuous transformation of research projects into ethical issues through the publication of papers and reports such as the ones the Nano2Life ethics board produced, and, consequently, continuous discussions about the European principles. Second, the integration also implied that the European scientists were trained to identify the ethical issues. This had a very practical dimension for scientists, through the mechanism of the ethical review process, which could be used as a tool to check whether scientists would use the code of conduct, acknowledge the EGE nanomedicine *Opinion*, implement safety measures in laboratories, and define the long-term objectives of their project in terms that would be consistent with European principles. Hence, a research project aiming to develop the "nontherapeutic enhancement of people" (e.g., brain implants meant to stimulate the cognitive capacities) would have no chance of passing the ethical review process as such. It would have to add an ethics board in order to "monitor the project," address ethical issues as they arise, and organize training in ethics for scientists.

Whether the material "brain implants" developed in the projects would be different because of the integrated European ethics is then another question. The "early identification of issues" that the ethics board device is expected to ensure is explicitly connected to the "flexibility" it offers: rather than solidifying constraining choices in regulatory texts (which would, for instance, ban certain types of brain implants, or require additional risk assessment procedures for nano substances), the integration of ethics in research projects is a way to "explore," "dialogue," and "identify long-term issues" without introducing mandatory requirements. As members of Nano2Life ethics board explained:

Researchers working with these new technologies have the obligation to thoroughly consider such issues and consequences before they start and while they carry out their projects. To discuss possible ethical implications with ELSA (Ethical, Legal and Social Aspects) experts early on in a project may relieve the pressure for regulatory bodies to be proactive in response to the high speed of the development, because normally regulations are based on long-term learning and experience. (...) This will help to prevent ethically, socially and legally nonacceptable developments for the benefits of patients—and also the success of the European health economy. (Berger et al. 2008, 248)

Thus, the integration of ethical thinking in research projects gets around having to introduce legal requirements, or create new entities subject to stricter regulatory control, or even ban entities from the European research area (e.g., "brain implants" or "drug delivery devices"). Instead, it is meant to ensure that minimal safety measures are adopted, patients are informed, scientists are trained in ethics, and dialogues are undertaken with the general public. In so doing, the integration of ethics in research projects can operationalize European principles without hindering scientific development, and thereby solidify the integrated moral and political space of European research for yet another technological domain without enlarging the set of regulatory texts.

The initiatives undertaken within projects' ethics boards echo those of the numerous European projects funded to examine the "Ethical, Legal and

Social Aspects" (ELSA) of nanotechnology. With regard to the Nano2Life's ethics board, many of its members undertook close examinations of the ethical issues surrounding nanotechnology as well as operations meant to gather public opinion, or even to formulate "lay ethics," which would, thanks to " the absence of those working in nanoscience and industry" "allow (...) participants to define concerns and questions on their own terms and without being constrained by the more rigid formats associated with deliberation" (Davies, Macnaghten, and Kearnes 2009, 33).

No exhaustive description of the many ethics boards and ELSA projects is needed in order to identify a problematization of nanotechnology that is specifically European, and which, in turn, is a problematization of Europe itself. The central problem in the making of responsible futures for the development of nanotechnology in Europe is that of the appropriate way of defining common values and acceptable channels of public participation, and, eventually of the construction of the European polity. As such, and to reuse the expression introduced by the authors of the CTEKS report, nanotechnology appears as a true "European experiment," not only because it engaged important European resources for objectives that were different from the American science policy programs, but also because it problematized the very identity of Europe as a political entity. This experiment was not isolated: it paved the way for a more general transformation of the entire European research policy.

Redefining the European Research Policy

In a report written in 2013 for the Directorate-General for Research and Innovation, a group of scholars and officials who had worked on nanotechnology ELSA explicitly linked the nanotechnology experiments to the development and strengthening of "Responsible Research and Innovation" (RRI) (European Commission 2013). They contrasted the initiatives undertaken for nanotechnology with the case of GMO, read as an example of failure, and use them to propose ways of calling for RRI. According to them, RRI would allow the European institutions to ground technological choices on the expectations of the European public, and anticipate potential controversies.

Heralded by the DG (European Commission 2012), RRI remained a vague notion, which is often made explicit by scholars working at the boundary between academic research and intervention in science policy. Their interventions are called for by the Directorate-General for Research and Innovation itself. Thus, the director of the European Research Area spoke to experts during a 2011 workshop devoted to RRI in those terms:

"We need your help to define responsible research and innovation. After several years of research on the relation between science and society, we evidenced that we need to involve civil society very upstream to avoid misunderstanding and difficulties afterwards. ... We cannot guarantee the social acceptability for anything but the more we have dialogue the easier it is to understand the potential obstacles and to work on them" (qtd. in Owen, Macnaghten, and Stilgoe 2012).

The people to whom this call was addressed had been involved in the definition of nanotechnology as an experiment in the making of a democratic Europe. Some of them were university scholars, who saw in nanotechnology an "opportunity" for the social sciences to transform the European research policy (Macnaghten, Kearnes, and Wynne 2005) and engage in the development of "responsibility" as a characteristic of European science (Owen, Macnaghten, and Stilgoe 2012). Others were officials from the Directorate-General for Research and Innovation and had advocated for instruments such as the Code of Conduct (von Schomberg 2009). According to them, RRI would be "really European" if it managed to "radically transform" innovation processes by being receptive to public values and needs (Owen, Macnaghten, and Stilgoe 2012).

Whether or not innovation processes are or will be "radically transformed" is not my main interest here. I am much more interested in what RRI says about the desirable European future, and the way of constructing it. After the nanotechnology experiment (that is, a test for the transformation of the European science policy), technological development appears, in RRI, as a collective project within which controversies are anticipated, while responsibility is distributed thanks to nonconstraining instruments (such as codes of conduct and ethics review processes). This implies a close integration of the reflection about the objectives and practices of responsibility within the technical components of the European research policy.

This evolution is directly related to institutional changes within the Directorate-General for Research and Innovation of the European Commission. Hence, the DG's "Science and Society" unit became the "Science in Society"—itself a sign of "integration"—then the unit disappeared. The members of the former unit described this evolution as the restructuring of the European science policy around RRI: "The 'Science in Society' unit disappeared as such but ... Responsible Research and Innovation emerges as a governance concept and as a cross-cutting requirement."[33]

In the self-description of the members of the Science in Society unit, nanotechnology is mentioned as a crucial step, which makes the transition between attempts at bringing science closer to the European public in the early 2000s and the definition of RRI as a characteristic of the entire European science policy (ibid.). That RRI is presented, in the previous excerpt, as a "governance concept" becoming a "cross-cutting requirement" is a sign, in the technocratic language of the DG, that the considerations pertaining to the relationships between science and the European public are now extended to the whole European science policy. This is the outcome of the nanotechnology experiment, which makes moral and political integration reach a new step.

European Experiments

The fact that the problem of making responsible futures for nanotechnology in Europe is dealt with by emphasizing integration is not foreign to the American example discussed earlier. Insisting on the "dialogue" between nanoscientists and nanoethicists, and advocating for the integration of ethics at the heart of scientific research echoed the "embedded humanism" approach that is a component of real-time technology assessment. There are indeed links between the European and American sites examined in the previous pages. The coordinator of one of the European ELSA projects called Nanobio Raise was also conducting fieldwork in a Dutch laboratory as part of a CNS embedded humanism project, and the only U.S. ethicist who participated in the Nano2Life workshops devoted to the examination of nanobiotechnology's ethical issues was George Khushf, the American ethicist and proponent of an experimental form of ethics. But the multiple connections between the American and European integrated ethics should not hide their dissimilarities. As detailed in the previous section, the American integrated ethics of nanotechnology had to ensure the objectivity of its position by constructing a small-scale social scientific experiment through which it was possible to demonstrate the validity of real-time technology assessment and that attempted to intervene in the construction of material objects (recall the "safety by design approach"). In Europe, the integration was not meant to separate the expert work of social science from that of policymaking. Nor was it directed to the construction of material objects. Rather, the instruments introducing ethics in science policy programs and European ELSA projects were expected to ensure an exploration of the ethical issues of nanotechnology and provide recommendations to the European Commission in a way that would not bypass the principle of subsidiarity, but could, however, contribute to the construction of the European nanotechnology policy. The European agencements that ensured the responsibility of nanotechnology development used nonconstraining instruments intended to incite researchers and research institutions to act according to harmonized values. Accordingly, the problem of making nanotechnology futures responsible was answered in Europe in terms of the moral and political integration of the European research space thanks to the government of common values. As such, this problem is a window into more than just the ways of conceiving technological development. It also offers empirical insight for the study of European institutions as they struggle to make values part of what is governed at the European level and to reinvent both their methods of intervention and their very nature as would-be democratic bodies in search for legitimacy.

Engaging

6 Mobilizing against, Mobilizing within Nanotechnology

Comparing Two Forms of Social Mobilization

The previous chapters have identified various problematizations of nano-technology, which have been described through the analysis of agencements. We encountered agencements that represent nanotechnology in museums; agencements based on technologies of democracy; agencements that define regulatory categories; and agencements that make people, objects, and organizations "responsible." How then to reconstruct consistent democratic spaces from the variety of these descriptions? This question has two interconnected levels. First, one can seek to connect the descriptions to each other and let consistent spaces emerge, "consistent" in that they would be characterized by common problematizations. Second, one can then inquire into the form of critique thus made possible. So far, I have argued that each of the problematizations of nanotechnology I described also realized a democratic construction by organizing oppositions. How is it then possible to ground a critique of democratic societies from this analysis? Is there a path toward "democratization" once the nature and effect of democratic values are embedded in particular problematizations (and consequently, when there is no stable external basis on which critique could be grounded)?

The last chapter of this book will examine these questions in details, by connecting various empirical descriptions presented elsewhere in the chapters and by discussing the theoretical perspective the case of nanotechnology helps develop for the analysis of democracy. Before coming to this point, this chapter proposes a discussion of two French civil society organizations (let us use the expression "civil society organization" here, which is not, as we will see, entirely appropriate). This detour is not anecdotal. First, it will provide empirical materials useful to complement the previous descriptions, especially about the state experiment undertaken with nanotechnology in France (cf. chapters 3 and 4). Second, and perhaps more importantly, these two organizations are caught in the same quandary as any scholar interested in the questions introduced in this book. They attempt to describe nanotechnology as a heterogeneous assemblage of objects, futures, concerns, and publics. They reflect on the path for democratization. They refuse to engage uncritically in public engagement devices but seek to actively undertake them as objects of reflection and intervention. They problematize nanotechnology by accounting for the problems it raises while also contributing to make it a public problem.

As in the preceding chapters, this chapter builds on comparative analysis. But whereas the previous chapters compare sites of problematization in different countries, I compare here the modes of social mobilization of two organizations in the same country, France. The comparison is relevant, I contend, since it helps illustrate two forms of engagement in nanotechnology that mix description and intervention. The narratives presented in this chapter will thus help me clarify the nature of the scholarly intervention in nanotechnology, and, more generally, in issues related to democratic practices.

The previous chapters have already provided numerous examples of social movements that are involved in controversies related to nanotechnology. Recall the fight of ICTA (in the United States) or the EEB (in Europe) for the recognition of nano substances as "new" chemicals (see chapter 4). In the meantime, we also encountered numerous devices expected to speak for the public, whether when creating debating or deliberating citizens (chapters 2 and 3) or through the "monitoring of public opinion" (chapter 2). The cases presented in this chapter will provide illustrations of different processes, whereby civil society organizations are engaged in nanotechnology as a whole, and question the way of maintaining, or not, a distance to it.

I will begin the description of these groups with an analysis of nanotechnology debates in Grenoble. Grenoble, a city in the French Alps and a major hub for nanotechnology research, is the place where they interacted for the first time, and where many of the actors encountered so far first met. In Grenoble, the connections are numerous among the making of nanotechnology objects, the definition of nanotechnology development programs, the management of its related concerns, and the mobilization of its publics. The Grenoble case will ground the discussion on the forms of mobilization proposed by a group of anti-nanotechnology activists, and an NGO engaged for the "democratization of nanotechnology."

In Grenoble: Introducing Social Mobilization on Nanotechnology

Nanotechnology and the Grenoble Model

Nanotechnology research in Grenoble has developed through tight collaborations among industrial, scientific, and administrative actors. Indeed, scientific and industrial research in Grenoble has received strong support from local administrations. The Grenoble city council, the Grenoble metropolitan area council (nicknamed *La Métro*), and the Rhone-Alpes region have been providing funding for scientific projects. Collaborations between public and private institutions for microelectronics and nanoelectronics R&D were launched in the early 1980s (Robinson, Rip, and Mangematin 2006). They reached a higher level of development in the mid-2000s with the Minatec Research Centre, launched in 2002 and officially opened in 2006, which aimed to "become Europe's top centre for innovation and expertise in micro and nanotechnology."[1] Another core research area in Grenoble is nanobiotechnology. The Nanobio project was launched in 2001 by the Commissariat à l'Energie Atomique (CEA), which has one of its major laboratories in Grenoble, and the Joseph-Fourier University, with the financial support of local authorities. Nanobio, which was conceived as a part of the European Network Nano2Life (see chapter 1), brought together engineers, physicists, and biologists within a broad portfolio of activities, from bio-imaging and bio-detection to surface chemistry, "at the interface of biology and micro and nanotechnology."[2]

The tight connections among industries, local administrations, and public research had another dimension, particularly as local officials insisted on the strategic objectives of scientific research in the Grenoble area. For them, science—and in particular nanotechnology—was to be developed for its economic value. When city officials explain their support for nanotechnology, they stress the economic dimension of these research programs. The Nanobio project seeks to "stimulate company creation and technology transfer"[3], and Minatec research center is part of Minalogic, one of the *Pôles de Compétitivité* created by the French government in 2004 in order to foster university-industry relationships.

The interconnections of public and private actors, scientific institutions and industries, for the sake of techno-economic development did not arise in Grenoble with nanotechnology. It followed a path opened after World War II, which embedded research in the physical sciences in a dense network of collaborations between scientific and industrial actors as well as with the city public administration (Caron 2000; Pestre 1990). The reference to Louis Néel, the Grenoble-based physicist who received the Nobel

Prize in 1970 for his work on the magnetic properties of matter, allows local actors to stress the tradition and continuity of the "Grenoble model." In his speech marking the inauguration of Minatec, the president of the local council of Isère explained how the new research center was being launched in the spirit of Louis Néel: "Professor Louis Néel ... said: 'I wish to develop a multi-disciplinary institution and link it to the whole set of regional industrial activities, as well as to the university and the CNRS.' It is this vision that has inspired the Minatec innovation centre, and that is why Minatec is situated on a square named after the 1970 physics Nobel Prize winner.[4]

Hence, for the Grenoble actors, the initiatives of CEA officials who pushed for the development of nanotechnology by administrative bodies, industrial firms, and public research institutions were pursuing the "Grenoble model" on yet another scale—that of the "technology of the future."[5] The "model" has practical meaning for the administration of scientific research, as a CEA official explained to me during an interview: "The local administrations are strongly involved in the emerging scientific issues. Everyone knows each other here, in the industries, in the labs, in the city council. ... So decisions are made quickly, and engagements are respected. ... Grenoble is quite unique in this respect.[6]

This specificity of the Grenoble model has been used as a reference for national policy for the development of innovation clusters. For instance, reports commissioned by the French government to identify processes ensuring national competitiveness referred to Grenoble as the "good example," a place where the close connections among university, industry, and local administrations were able to produce scientific knowledge and transfer it.[7] In the discourse favored by local officials, the Grenoble model is both integrated and comprehensive, the result of the past and a marker for the future. It is both a condition for the success of scientific projects and the reason for the continuation of scientific activity in Grenoble.

The Grenoble model conflates the technical contents of the various S&T fields with the organizational aspects that render multidisciplinary connections possible. Research activities in the Grenoble area associate nanosciences, basic technological research, industrial R&D, and also expertise in software technologies, biotechnologies, and energy microsources. At the level of the laboratory, it means that the institutional arrangements, the scientific instruments, the research projects, and the work status of the researchers and engineers are redefined according to the need of the complete innovation system. Researchers in nanoelectronics laboratories share instruments across scientific disciplines and institutional boundaries, thereby redefining the nature of their projects in terms of both "fundamental" and "applied" research (Hubert 2007). Research in nanobiotechnology led to bringing patients previously cared for in the local hospital to CEA buildings, where, within a project called Clinatec, physicians, biologists, and physicists could experiment with nanomaterials-based cerebral probes.

The recomposition of institutional, disciplinary, and cognitive boundaries in the Grenoble area has been accompanied by a growing concern for the management of risks and the interrogation about potential ethical issues. François Berger, the promoter of Clinatec, sat in the Nano2life ethics board, and developed a constant preoccupation for the "ethical questions of nanotechnology research" (see chapter 5). CEA is a major partner in successive programs for the study of the health risks of nanoparticles. Regular academic conferences called Nanosafe are organized at Minatec. Employees of CEA participate in national and international discussions about the standardization of nano substances. CEA's lead occupational physicist is a member of the French delegation to ISO, and an active participant in the nano-responsible initiative (see chapter 4). In Grenoble, the concern for the public is also visible. Chapter 2 described the numerous initiatives of the local science center, and its involvement in the display and practice of the "public debate" about nanotechnology.

Against the Grenoble Model

In June 2006, the Minatec research center was officially inaugurated. President Jacques Chirac was expected, as well as ministers and multiple representatives of French and European research institutions. Yet what should have been the symbol of the success of the Grenoble model was disrupted by a demonstration on the streets of the city. Allegedly the first anti-nanotechnology march in the world,[8] the demonstration, which gathered about a thousand people, had been announced on the website of the *Oppositions Grenobloises aux Nécrotechnologies* (OGN)—which had caused, according to the activists, the defection of the President and eventually the shift of the inauguration day from June 1 to the following day.[9] The OGN demonstration was only a culmination of a series of actions opposed to nanotechnology, which had taken various forms in the Grenoble area. Activists had already organized various counterevents in bars comprising movie projections and discussions about nanotechnology. Over the 2000s, the contestation of nanotechnology became visible in Grenoble as "no nano" mottos appeared on the city's walls. These actions were pursued at the national level during the national public debate on nanotechnology, when activists

interrupted public meetings (see chapter 3). But the most important production of the anti-nanotechnology activists in Grenoble was by far the
writing of texts, in which anonymous authors would describe the tight connections among the Grenoble scientific, industrial, and administrative
actors engaged in the promotion of nanotechnology.

At the origin of the contestation of nanotechnology was a group called
"Pièces et Main d'Oeuvre" (PMO, or Parts and Labor), which defined nanotechnology as a "necrotechnology," that is to say a technology that has to
do with "death" (Greek: necros).[10] Indeed, the activists described scientific
research in Grenoble as part of a global program of control over nature and
human beings. PMO targeted the blurring of boundaries that nanotechnology produced. By merging biology and physics in the making of hybrid
objects, such as diagnostic tools and brain implants, nanotechnology, so
the activists argued, was a threat to the integrity of the human body, and
a potential provider of devices controlling human beings; by associating
fundamental and applied research, academic and industrial research for
the sake of economic development backed by public and private actors, it
destroyed the autonomy of science (including social science), and subsumed the public good to economic interests.

PMO is composed of no more than a few people, who are joined by
other activists in planning activities, writing texts, or demonstrating.
Hence, speaking of the "anti-nanotechnology activists" should not suggest
that they form a consistent social movement. They are mostly a collection
of people coming from various backgrounds, most of them being loosely
associated with various activist groups. Consider the trajectory of a member
of OGN, interviewed in a radio program: "I was trained as an engineer. I
worked for a big company. After a few years of this work, I was fed up with
dissociating my professional consciousness and my moral consciousness,
(...) I felt a dissonance between my principles, which lean toward ecology
and democracy, and my work. I decided to quit, for a life with much less
money but also many more friends and much more political concerns.
Since then, I gravitate among the opponents of the race for high-tech. Not
only in this group though."[11]

Like this person, the activists I interacted with (in interviews or meetings) were mostly educated people, who had decided to engage "against
technology." What an engagement "against (nano)technology" means will
be explored at length in the following pages. At this stage, it suffices to say
that the activists' description of the Grenoble model is quite different from
that of public officials. For the activists, the Grenoble model is not a success
story in terms of technological and economic development, but rather an

illustration of the increasing domination of market interests without public legitimization, eventually resulting in the weakening of democratic processes of decision making. A symbolic figure like Louis Néel is thus deconstructed as a representative of unacceptable contacts among basic research and military and economic interests.[12] For the activists, nanotechnology research is a manifestation of another type of convergence, that of political, scientific, military, and economic interests, which leads to decisions based on military or market interests, and, therefore, opposed to the general interest. Decisions in Grenoble, the activists claim, are made by a small group of people without prior consideration of citizens' interests. Officials and scientists constitute what the activists call the "techno-gratin" (the "techno-upper crust"), in other words, a small elite group whose members have close ties to one another. The case of the mayor of Grenoble, a former engineer in CEA and founder of a spinoff research center, is often used to illustrate this situation. This criticism is reinforced by the connection drawn by activists between nanotechnology research and local events apparently not directly connected to technology itself. For instance, the activists' definition of the Grenoble model includes references to past corruption scandals that involved high-ranking local officials.[13] Another example is the arrest of demonstrators by the police during the demonstration against Minatec in June 2006: this was interpreted as an attempt to enforce decisions about technology, as was the police intervention during the inauguration of the Grenoble nanotechnology exhibit (see chapter 2). As such, it was seen as another manifestation of the program of control inseparable from scientific research.

"Public Dialogue" as Another Site for Oppositions

The opposition to scientific projects was not ignored by local officials. The local councils commissioned various events that were variably described as "dialogues," "debates," or "forums." At the initiative of a councilor of a minority group, La Métro commissioned a report to a group of STS scholars in 2005. They were asked to work on nanotechnology and the local democracy in the Grenoble area (Joly et al. 2005).[14] The report made the importance of the Grenoble model explicit, and recommended that participatory mechanisms be put in place. It recommended in particular the organization of a citizen conference—which was ridiculed by PMO, as a lame attempt to display a fake concern for democracy while continuing supporting nanotechnology development.[15]

No citizen conference was organized, but La Métro commissioned a series of public debates called NanoViv, which a civil society organization named

Vivagora organized. When it intervened in Grenoble, it had been working on nanotechnology for a couple of years. It had organized a series of public meetings in Paris about nanotechnology, which were meant to "expose the opposing views," "confronting the arguments," and eventually "come up with recommendations" for a "more democratic, more transparent, more inclusive governance of nanotechnology."[16] The same model was followed in the organization of NanoViv.

In Grenoble, Vivagora and PMO directly opposed each other. As the director of Vivagora, Dorothée Benoît-Browaeys asked PMO to participate in the meetings she was organizing,[17] the activists released texts in which they explained that La Métro was "trying to recruit them"—and that they would not participate in this "parody of public debate."[18] For them, any public debate could be nothing but a component of the global nanotechnology program, which was to be mobilized against. Vivagora had no better luck with the administrative and scientific officials: they were "asked to participate in the meetings"[19] but no official acknowledgment of the recommendations was produced at the end of NanoViv. These recommendations were general, and mostly targeted the "lack of transparency" in nanotechnology research in the Grenoble area as they were advocating regular discussions with civil society organizations. They were not well received among Grenoble officials. Years after the Grenoble debates, scientists and officials who had participated in them would still regularly tell me that "Vivagora had made up the recommendations."[20]

For both activists and Grenoble officials, the intervention of Vivagora was to be criticized. For the former, it intervened in the very making of nanotechnology policy (of which dialogue was a central component to ensure at best the "management of impacts" of an unquestioned technology program, at worst the "enrollment" of passive populations) and thus could not pretend to observe nanotechnology from the neutral position PMO contended to occupy. For the latter, Vivagora was involved in ways that went far beyond what it was paid for (i.e., organizing public discussions through which, as a city councilor said to me during an interview, "nanotechnology and its impacts could be presented in a manner that would take the heat out of the debate"[21]). Vivagora's interventions could have been acceptable if it had been an expert in public debate (like the CNDP [National Commission for Public Debate] experts encountered in chapter 3), but its inability to solidify a participatory procedure, to make it independent from the object being discussed, and to eventually produce uncontested results, made it a target for the critique of the Grenoble actors.[22]

Two Forms of Social Mobilization in Grenoble

Grenoble is a site where the components of nanotechnology stand out clearly. Many of the actors mentioned in the previous chapters converge in Grenoble, as the local construction of development projects is connected to the global construction of nanotechnology. Ethical concerns are voiced, the safety of nano substances is discussed, and nanotechnology's publics are engaged. It is a place where the development of nanotechnology as an entity gathering objects, futures, concerns, and publics is undertaken in a visible way. In Grenoble, one cannot reduce nanotechnology to a set of unconnected applications, to a problem of risk management, or to anticipatory visions. Accordingly, it is the place where social mobilization considers nanotechnology as a global program to be targeted, in ways that differ from those of the civil society organizations we encountered in the previous chapters. Opposing a model sustained by local officials, which contends that nanotechnology should be developed for the sake of local economic development while citizens should recognize the validity of expert knowledge and witnesses the management of the risks of each individual nanotechnology product, anti-nanotechnology activists define nanotechnology as a global program of control over nature and human beings, against which citizens need to engage. This latter proposition implies that activists refuse to engage in participatory activities, in order to critique them—as components of the global nanotechnology program they oppose. In arguing for the "democratization of nanotechnology," Vivagora takes a different stance that contends that nanotechnology should be open to collective discussions, which the organization would be in charge of setting up. This seems to imply that the mobilization is based on procedures meant to transform nanotechnology into a series of projects to be constructed by engaged actors, whether experts, administrators, or interested citizens—a proposition that was not well received in Grenoble.

PMO and Vivagora propose original forms of engagement in nanotechnology. They provide examples of practices that do not always follow the distinction between "invited" and "uninvited" participation in public debates, and that propose contrasted modes of critical engagement in or against nanotechnology. The sections that follow describe these two cases.

Mobilizing against Nanotechnology

Demonstrations without a Social Movement

For the local officials and research administrative actors, PMO was "not representative." As a member of the Grenoble city council said, "An overwhelming majority of people supports the development projects. These people are a tiny fraction of Grenoble inhabitants. ... They're not representative."[23]

The same type of critique ("not representative") was repeatedly heard during the CNDP national public debate on nanotechnology. The "nonrepresentativeness" was a recurring critique of the president of the team in charge of the debate, who would present the number of connections to the website and the number of participants in the public meetings, then compare these figures with the "reduced numbers" of activists.[24] Whether the quantitative arguments hold true or not,[25] the critique misses the point. That anti-nano activists are not numerous is not what matters, since the type of critique they articulate cannot be differentiated from the particular format of action they propose, which is not expected to "represent" particular stakes or social groups.

This particular form of intervention is based on the anonymity of the critical voice. Unlike an organization claiming to represent certain people or issues, PMO's voice is anonymous. This is not trivial, as constant complaints are heard on the part of officials and scientists, who blame the opponents for not "playing the game of democracy" by "refusing to appear as persons with a name."[26] That democracy is at stake is certainly the case. This is not because PMO would not follow the "rules of democracy," but because it proposes a model of citizenship in the democratic society based on "critical inquiry" (*enquête critique*) performed by an individual and neutral "simple citizen" (*simple citoyen*), situated outside of the making of political, economic, and technical decisions. As one of the members of PMO explained, "Refusing to display our identities was deliberate. There are so many people who want to be known ['se faire un nom']. We are not here to build our notoriety; we do not want to be celebrities on these topics. (...). There are three types of authority: scientific, political and related to the media. We have wanted to act out of all that. Judge us on what we do ['sur pièces'], on the texts we write, which are all sourced."[27]

Constituting PMO into a social movement would have meant that the group would have fought for a particular stake, whereas it precisely sought to avoid being part of the negotiation game. Rather, it preferred to conduct critical inquiry, to develop a fine-grained assessment of the interests of the involved actors. Such a position is not foreign to social science. It echoes Bourdieu's perspective on sociology, which objectifies social categories in order to conduct the (social) scientific demonstration. Bourdieu's sociology relies on the ability to maintain a position of exteriority, which, as sociologists know, can be challenging. Reflexivity, then, as a means for the "objectification of the process of objectification" (Bourdieu 1980), is expected to allow the sociologist to situate her position. For the Grenoble activists, the problem of exteriority was solved not by the recourse to reflexivity but by anonymity—a necessary requirement for the critique of PMO to be articulated.

This directly impacted the form of mobilization, as the contestation of nanotechnology could not be constituted into a "social movement." PMO itself is mainly composed of a handful of people, while friends maintain the website, and friends of friends organize meetings and debates. A student at Grenoble University who had written a humorous (and critical) account of one of the NanoViv public meetings thus explained during an interview: "Yes, we all know each other ... I had a friend who knew Y. [one of PMO's members]. I went to a few meetings in Grenoble. I had a good idea of what is happening in Grenoble. In this case, I found it fun to write a short piece (...) This is often how it works. Someone takes the initiative to write something, and then we circulate it. There is not much more organization."[28]

This does not mean that no organization exists at all. There are indeed multiple connections among people interested in the contestation. Information is exchanged, informally as the previous quote illustrates, or through alternative web media platforms,[29] and ad hoc groups of people are constituted when preparing particular demonstrations (e.g., the OGN group, for "Opposition Grenobloise aux Nécrotechnologies," which organized the demonstration against the inauguration of Minatec).

The anonymous position means that the form of PMO's demonstration is not performed as a public display of a particular stake or interest. As Andrew Barry suggests (Barry 1999), one can use the term "demonstration" to point to two operations: the performance of public proofs and the social event expected to make issues public. But contrary to Barry's examples, which deal with a case of mobilization against a planned highway in the British countryside, where the demonstration was about the connection between the people and the land, the demonstration that the Grenoble anti-nanotechnology activists proposed was not directed to one particular issue, but multiplied into a wide range of public proofs. PMO activists and their friends in the Grenoble area have worked hard to render visible the multiple connections that constitute the network of people in scientific, industrial, and administrative spheres that regularly interact and allow the Grenoble model to be sustained. One of PMO's most distributed productions is a graph that displays the multiple links among the officials in the local administrative bodies, the industries, and the management of scientific research. Connected to this representation of the control of the local decision-making process by a small group of people is the demonstration of the physical transformation of the city of Grenoble. The occupation of a crane during Minatec construction work is a telling example of such demonstration. Another type of demonstration is based on ironic and humorous interventions: in 2007, activists distributed a fake version of the information magazine of the local administrative council, announcing the end of nanotechnology programs in Grenoble.

Ultimately, the demonstrations performed by anonymous "simple citizens" do not aim to constitute a social movement and argue publicly for the validity of a particular stake or interest. Rather, they aim to render the critical gaze directed toward nanotechnology immediately visible. In doing so, they act as devices through which spectators of the demonstration can be turned into critical citizens, potentially contributing to the critical inquiry. That these demonstrations are performed by anonymous actors is important, for that matter: it is a way of creating a parallel public space, "public" in the sense that it belongs to every citizen and not those particularly affected by a given problem. This parallel public space operates outside the scope of the official one. It is composed of websites, independent media, and places in the Grenoble area (and, during the CNDP debate, all over the country) where public meetings were held and activists discussed nanotechnology.

The parallel public space does not operate with publics other than simple citizens. In the model of the critical inquiry that the anti-nanotechnology activists propose, the nature of the "public" of nanotechnology is indeed twofold. On the one hand, the "official public" is part of nanotechnology programs, which comprise public meetings, dialogues and forums, and can be probed by measures of public opinion. On the other hand, the simple citizen is expected to put nanotechnology at a distance in order to demonstrate the interests behind its development and the noxious links on which it relies. Thus, the "public" of nanotechnology and its "problem" are conceived at two separate levels by PMO. On the one hand, the "problems" of the "impacts" of nanotechnology and their associated publics fit perfectly well within the global nanotechnology program that is to be rejected. On the other hand, nanotechnology *as a global program* comprising not only technical objects and future developments, but also publics and concerns, is a problem for which individual simple citizens need to mobilize and that needs to be discussed and acted upon through spectacular demonstrations performed in the parallel public space, but visible from the official one.

Thus, PMO's critique attempts to reconstruct a space for citizen intervention about technological issues. This space is both material and abstract. It is material in that it relies on the many places where activists meet and perform spectacular demonstrations. It is abstract in that it is from there that the simple citizen produces her critique. It is in this space that the citizen may become "simple," detached from any private interest in ways that resonate with a Rousseau-ist theory of democracy. Rousseau's *Social Contract* can be read as a theory of the general interest, construed as a condition for social stability, and requiring that private interests are subsumed into it. This political philosophy makes citizens "simple," in that they are supposed to gain equality of rights by renouncing their particularities. This figure of citizenship, central in the development of the French democracy (Rosanvallon 1992, 2011b), is remobilized by PMO as a way of contesting what transforms citizens who could be "simple" by turning them into "debating citizens," by making economic interests the matter of national and local research policies, and by developing technologies that might act on the very identity of political subjects.

Constructing Distance to Nanotechnology

The refusal of nanotechnology that PMO proposes is more complicated than a request for a moratorium, as some civil society organizations advocate,[30] and which requires boundary work in order to distinguish what is nano and what is not. PMO refuses to enter the process of defining nanotechnology from within, and blames what shifts and blurs boundaries: scientific developments that are at the same time economic development programs, social scientists who intervene in the conduct of nanotechnology research, and opponents who negotiate with industrialists about norms and standards. Accordingly, PMO is not interested in debates about the definition of nanotechnology objects in standardization and regulatory institutions, nor in discussions within science policy offices about how to make nanotechnology futures "responsible" (see chapters 4 and 5). An analysis of the oppositions within ISO about how to define nanomaterials, for instance, would be considered unnecessary for PMO. Arguing that other definitions than the size-based ones are possible, albeit eliminated within the international standardization processes, would, for PMO, still contribute to the development of nanotechnology.

The challenge of PMO's social mobilization is to maintain the distance from which critique can be voiced. I experienced this directly when I first attempted to meet PMO activists in January 2007. We had several email exchanges (using an anonymous electronic address) before I could settle a meeting. The meeting happened at night in a low-key bar close to the Grenoble railway station, where I was asked to go and "be ready to be recognized by them." Two people eventually came to me, and it was only after another hour of discussion about the objective of my work and my tie to nanotechnology that I could start asking them questions. A young graduate student at that time, with no funding originating from nanotechnology programs, I needed to convince them that our meeting could be a contribution to my academic research without involving them in the entity they wanted to critique at a distance.

In other cases, PMO activists preferred turning down offers to voice their opinions alongside others'. A notable exception—and a challenge for PMO's critique—was the four-month debate organized in 2009–2010 by the CDNP at the initiative of the French government (introduced in chapter 3). What to do with the CNDP debate was indeed an issue for the anti-nanotechnology activists. Not that participation alongside NGOs, industries, and government bodies was considered for a minute. The organizers did contact PMO and asked them to participate as an "official" member. But participating was of course not an option, for it would have meant that the activists would have entered the game of public discussions about nanotechnology. Yet maintaining the distance, in this case, could not be limited to a refusal to participate. For the national debate on nanotechnology was an opportunity not to be missed to perform spectacular demonstrations: demonstrations that the device was organized by the proponents of nanotechnology, that it was driven by the interests of nanotechnology development, that participation was not an acceptable way for the citizen to act, and, eventually, that the objective of "total inclusion" was absurd. The parallel public space had to interact with the official one.

Yet the CNDP device is meant to include as many forms of expression as possible, including the most critical ones, and the organizers are ready to adapt the procedure in order to look for diverse participants. This was not ignored by the activists when they discussed the format of their interventions. During a meeting in February 2010 in Paris, in which I sat, about fifty activists discussed the way in which they could intervene during the first Parisian meeting of the CNDP debate.[31] An issue that was debated was the opportunity to stress the negative aspects of nanotechnology in order to convince people to join the anti-nanotechnology movement. The proposition was not well received, since "negative aspects" like risks were already part of what was being discussed within the CNDP debate. Activists preferred targeting "worldviews" such as "the machine man in a machine world" (*l'homme machine dans un monde machine*). Yet even if "the machine man in a machine world" was accepted as the object of the critique, the forms of the demonstration were not given, and could potentially be harmful for the activists, since it could also render explicit *to the organizers of the public debate* a position that could then be included as "the opinion of the activists" alongside that of the participants in the debate. The activists did not ignore this, as they reflected during the February 2010 meeting on the possible means they could use to perform the demonstration. They considered "taking the mic" (*prendre le micro*), but eliminated the option because of the risks that the activists would take to appear "just like the other participants." "Shouting" was eliminated for the same reasons. Discussions about the banners lasted for a few minutes. The activists had crafted a series of mottos, and published them on the website. But the problem was still present. For instance, one of the mottos targeted the financial interests of the organizers of the public debate. At the Rennes meeting, the organizer of the debate could directly answer this critique, and start a discussion about the wages of the members of the organizing commission members ("it's a good thing you ask, because we are not paid," said one of the organizers). Hence, every formulation of argument within the perimeter of the public meeting, under one form or another, was to be considered within the dialogue device and then be captured in it. The activists' interventions eventually did not even try to convey arguments but were meant to render the conduct of the debate impossible: activists would blow whistles, shout unformed words, and refuse to talk to whoever was asking them questions. This was the price to pay in order to stabilize the distance to an inclusive device absorbing every argument, while in the meantime conducting spectacular demonstrations. Even so, the organizers of the debate were able to devote a large part of their final report to the interventions of the activists. The organizers presented the opponents as participants who had been able to shift the debate to questions that the organizers considered more interesting ("opportunity"). While criticizing their refusal of "dialogue," they considered the positive side of the activists' interventions, which "increased the visibility of the debate" and raised the questions of "the society we want" and of "governance."[32] The intervention of PMO within the CNDP debate can be seen as a trial for PMO's mode of critical intervention. While PMO succeeded in interrupting public meetings, the conduct of spectacular demonstrations within the CNDP debate raised the issue of the practical construction of the distance from which critique was possible.

Extending Critical Inquiry

The crucial role of the CNDP debate for PMO is also visible when one considers the extension of critical inquiry beyond nanotechnology. The debate was an opportunity to gather various people under the banner of the "converging fights"—an expression regularly used by the activists, and which mirrors the "converging technologies" they are acting against. During the CNDP debate, the activists who intervened came from various cities in France. Some of them traveled across the country, going from one debate to another. Some had been active in the critique of nuclear energy, others in the anti-GMO movements: in all these cases, what was at stake was the possibility for the simple citizen to perform a critique, and what was targeted was the democratic organization.

As CNDP was organizing its last public meeting, about two hundred activists coming from various locations and with diverse experiences in anti-technology activism gathered in Paris to discuss the actions to undertake after the CNDP debate.[33] They compared the experiences they had had with various environmental and green political parties in order to examine how the "converging fights" could develop. While participants talked about the cooperation they had had with *Confédération Paysanne* (an anti-GMO farmer union) or *Sortir du Nucléaire* (the main French anti-nuclear group), others (most notably the most active PMO members) warned against the risks of entering "a dynamic of negotiation rather than contestation." They used the illustration of the anti-nanotechnology contestation to argue for the pursuit of critical inquiry and the refusal to constitute a social movement that would engage in participatory mechanisms. At stake was again the stability of the distance from which critical inquiry could be performed, and which was crucial for the construction of the "converging fights."

As the group is not organized and does not argue for a given interest, the nature of its critical intervention is constantly reopened, and the ways of producing the adequate distance to technologies of democracy such as the CNDP debate are not pre-given. This results in narratives about past choices, past interventions that helped clarify the mode of action but ended up raising new questions. That the written production of PMO is abundant is not anecdotal for that matter. In addition to the many texts circulated online and within activist networks, it also includes several books published by independent publishers (PMO 2008, 2009). These productions present the results of critical inquiry while also reflexively re-narrating actions undertaken in the past and discussing their contribution to critical inquiry. Consider, for instance, the case of activism against nuclear

energy, in which the instigators of PMO were actively involved. While criticizing the civil society organizations that chose to participate in public
dialogues about nanotechnology, PMO often refers to previous events
related to the critique of nuclear energy, sometimes quite distant in the
past, during which differences in the conduct of the critique of technology
had been visible. For instance, a demonstration conducted against a
nuclear project in Creys-Malville in 1977 led to lengthy debates among the
activists about whether or not violence was acceptable. Writing in 2005
about this event, the simple citizen considered it marking the separation
between the environmental movements that were willing to negotiate
with public bodies and private actors, and radical activists (which were, for
the author, the only ones able to perform critical inquiry).[34] The Malville
demonstration was yet another site where the practice of critical inquiry
was at stake. It recomposed the spectrum of antinuclear activism in ways
that, according to the PMO narrator, continue to have consequences in the
2000s. Described in the mid-2000s by PMO, it became a component of the
parallel public space of radical critique, and another moment of trial for
the stability of critical distance.

How to Make PMO a Research Topic?

While PMO constructs a parallel public space from which critical inquiry
can be undertaken, it integrates social science in its critique. Thus, PMO
produces numerous critical accounts of social science, whether commenting on STS scholars advising *La Métro* to organize a citizen conference,[35] or
discussing the concepts of "technical democracy"—which does not question, according to the activists, the very logic of technologic development.[36]
We have already encountered the intervention of social science in nanotechnology programs in the previous chapters, as I described social scientists involved in the design of science exhibits (chapter 2), in the organization
of technologies of democracy (chapter 3), and in the definition of "responsible" futures for nanotechnology (chapter 5). But the intervention of PMO
invites one to theorize further the position from which such a description
is even possible. Indeed, including the anti-nanotechnology movement
within problematizations of nanotechnology described at a distance would
replicate PMO's approach. But rather than adopting PMO's exteriority solution, one can contrast PMO's production of exteriority with other ways of
producing critical distance, particularly those in which the analyst himself
is engaged. If we follow this approach, PMO is less an entity to be put at
a distance than a group of people raising issues similar to this book's, of
the good description of the issue at stake, of engagement, and of the

construction of democratic orders. As such, the practical issues that PMO faces in maintaining its exteriority within inclusive agencements, and its refusal to pursue its critical inquiry in places such as standardization organizations and science policy offices show that the descriptions produced by PMO might miss important sites where nanotechnology is problematized. The next section contrasts PMO's production of exteriority with the multiple distances produced by another organization, Vivagora.

Mobilizing within Nanotechnology

Engaging for the Democratization of Science

Vivagora is a small organization, which never reached more than a couple hundred members and a small group of employees. But its involvement in the discussions about nanotechnology in France was far more significant than what its size would suggest. Vivagora is an organizer of "public dialogues" such as, in Grenoble, the NanoViv debate series. It is tempting to think of Vivagora as a "mediator" that would "not take part" in the discussions it helps organize. This would separate nanotechnology from its treatment within a public debate arena, organized by a specialist, in this case Vivagora. This, however, does not account for the activities of the organization: Vivagora is neither a "stakeholder" nor an "advocate of public debate" using participatory instruments independently of the issue at stake; the case of Vivagora is significant because of the impossibility of delineating a priori boundaries for nanotechnology. The organization indeed became part of nanotechnology programs, allied to administrative actors involved in the making of science policy, while, at the same time, constantly reflecting on the specificity of its position.

Another characteristic of Vivagora is indeed to be self-reflecting. Members of the organization and employees regularly gather to discuss the objectives of their group. During the board meetings of the organization, members invite external speakers, express their wishes about the future of the organization, and discuss its "identity." "We have an identity crisis" is a sentence I heard many times during my exchanges with Vivagora members and employees.[37] That the organization was so concerned about its identity, its "values" and "objectives" is connected to the nature of its commitment to the "democratization of technology." As will be seen in this section, Vivagora's concern for democratization could not be separated from an engagement in the actual production of nanotechnology objects and policy instruments. This makes the case of Vivagora different from that of the experts of participatory procedures encountered in chapter 3.

As I was participating in numerous meetings about the definition of nanomaterials, the identification of ethical issues, or the design of science exhibits, I had multiple interactions with Vivagora, whose members were present in all the fieldwork I worked on. Vivagora collaborated with the Grenoble science center in the organization of local public meetings in connection with its nano exhibit (see chapter 2). It participated in the CNDP debate, at first a supporter of the process, and then turning more critical. The organization was involved in discussions at AFNOR about standardization and the nano-responsible project (see chapter 5), then later joined the European Environmental Bureau, and signed a declaration about the "principles for the oversight of nanotechnology" prepared by the American ICTA (see chapter 4). More than from external interactions with the organization, a significant part of my knowledge about Vivagora stems from my active involvement in it. As I began working on the nanotechnology debates in Grenoble, and, more generally, on the assemblage of nanotechnology in democracy, I interviewed members and employees of Vivagora and observed some of the events they were organizing. I was increasingly involved in their activities and entered a process of ongoing negotiations about my role and relationships with the organization.

The Experience of the Nanoforum: From an Expertise on Public Debate to the Construction of Publics and Problems

Vivagora was created by science journalists and conceived at first as an organization that provided information about controversial topics related to emerging technologies. This was the spirit of several of its initiatives in the mid-2000s, when the organization organized public meetings during which the public's concerns related to nanotechnology were discussed by actors in the field. But the many criticisms it encountered in Grenoble forced the organization to rethink its mode of intervention and the objective of its actions. The Grenoble experience was interpreted by Vivagora members as the demonstration that the external position from which one could describe the various positions related to nanotechnology was a fantasy. "We're included," the president of Vivagora repeatedly said. Whereas PMO responded to the quandary caused by nanotechnology's inclusive character with the anonymity of a simple citizen performing critical inquiry at a distance, Vivagora chose to rethink what it meant to mobilize on the "democratization" of technological choices. Rather than picturing technological issues at a distance, as if it could maintain an external position, the organization would engage actively into them.

This shift manifested itself in the initiatives Vivagora undertook after the Grenoble experience. One of the most visible was the Nanoforum, yet another series of public meetings about nanotechnology, albeit grounded on a different approach than the Grenoble events. The Nanoforum was organized in Paris, and lasted from 2007 to 2009. A first group of meetings focused on various industrial domains of application of nanotechnology (e.g., construction and cosmetics). They were examined through the participation of invited industrialists, civil servants, and representatives of NGOs. Another group of meetings was devoted to nanosilver and the modalities of its regulation within French and European law. Rather than delving into the details of the discussions during these meetings, it is more interesting for our concerns here to characterize the agencement that the Nanoforum resulted in, and the problematization of nanotechnology it entailed.

First, the Nanoforum was experimental in the sense that it made participants question their roles and responsibilities. For the civil servants involved, the scope of their engagement toward nanotechnology and its regulation was at stake: because they did not use a known technology of democracy (as the CNDP debate procedure), the modalities of action were not preestablished. For Vivagora, the form of social engagement for the democratization of nanotechnology (and technology in general) was directly at stake in the Nanoforum initiative. As the organization did not know in advance what it was looking for in putting together the device, it also explored the modalities of its engagement in public debate about nanotechnology through the Nanoforum. Eventually, the Nanoforum also called my own role as an engaged analyst into question. I was invited to participate in the organizing committee by Vivagora, as both a member of the organization and a nanotechnology "expert." While sympathetic to the objective of the Nanoforum I did not know in advance what my engagement was going to be.

The Nanoforum experiment was connected to others. Participants and organizers of the Nanoforum were active in the development of the AFNOR nano-responsibility standard (see chapter 4). The organizers of the Nanoforum wrote a contribution to the CNDP debate in which they explained that it was necessary to explore the uncertainties surrounding the risks and benefits of nanotechnology. As such, and it is the second aspect of the Nanoforum that I want to underline here, the initiative was a component of the state experiment in France that nanotechnology resulted in. The discussions undertaken within the Nanoforum dealt with choices engaging the state, and involved the civil servants in charge of nanotechnology

within the French public administration. The main initiator of the Nanoforum, William Dab, a former senior official at the ministry of health, saw
the initiative as a way of transforming the government of uncertainty
undertaken by the French state. In various public statements and publications, William Dab explained that the French state was ill equipped to deal
with uncertainty, because of the centralization of decision-making processes, too much reliance on technocratic expertise, and a refusal to
account for the politics of technological evolution (Dab and Salomon
2013). In contrast, the Nanoforum was for him an initiative that could be
seen as a "technology of trust," because its objective was to make uncertainties explicit, and to introduce dialogue across various components of
the French government before any rigid position had been crafted (Dab
2009). The Nanoforum was a component of a series of interventions performed under the eyes of European and international organizations that
targeted the regulation or standardization of nanomaterials in this context
of uncertainty (see chapters 3 and 4). It directly questioned the channels of
political representation beyond the election, as the CNDP debate would
soon do.[38]

There is a third experimental dimension within the Nanoforum, and it
is related to the political engagement of Vivagora. In its previous initiatives,
the organization had attempted to describe the various positions related to
nanotechnology—as science journalists would have described opposing
views on a technical question. Within the Nanoforum, Vivagora was not
only an organizer but also an active contributor. It intervened, for instance,
in the support given to local civil society groups in Grenoble, which
were invited to talk at the Nanoforum. This was a component of a broader
objective meant to "structure civil society" (an expression regularly used by
Vivagora's members), that is, encourage social actors to mobilize on nanotechnology. The example of the Nanoforum illustrates the trajectory of
Vivagora after the Paris and Grenoble debate series. Rather than stabilizing
a technology of democracy that could have been replicated independently
of the issue at stake, the organization preferred to set up ad hoc procedures
that could evolve according to the need of the participants and the questions that were raised.

This means that the external position, from which Vivagora could
have hoped to represent the various arguments exchanged within controversial situations was no longer possible. As a member of the board said
during a general assembly on March 4, 2009: "We have been in a process of
collective reflection on the vocation of Vivagora since 2008. At the beginning, we were above all interested [orienté vers] in public debate. Today, we

think it is better to intervene on innovation processes. We need to go beyond what we have already done [*aller plus loin*], we need to think about the ways in which social experiments are possible. How to launch new formats of public debates."

To "go beyond" could imply, as in the Nanoforum, work in common with social movements, administrative officials, and industrialists, in making nanotechnology a public concern, rather than organizing participatory procedures thanks to a procedural expertise. In any case, "social experiments" required transforming the forms of social mobilization.

Experimenting on Engagements and Distances

In addition to the publication of articles on its website and through its newsletter, which critically examined technological innovation and the evolution of national and European regulation (for instance, by commenting on public agency reports, and blaming the administration for not regulating nano substances), Vivagora was involved in 2008 in seventeen different projects.[39] About two thirds of them dealt with nanotechnology. Some of them consisted in the organization of public meetings on the model of the Grenoble public debates. Others were punctual events (such as an "Innovation and Responsibility" colloquium) or ad hoc processes such as the Nanoforum. Another group of activities were projects funded by public bodies intended to study participatory democracy on technology issues and experiment with forms of public dialogue. Other projects were regular events organized for industries (e.g., breakfast meetings to present the positions of various stakeholders to industrial actors), or partnerships with industries to experiment with "participatory design" (that is, industrial design involving representatives of civil society organizations). Eventually, Vivagora also participated in public committees where civil society organizations were invited to intervene. For instance, when the National Council for Consumption (CNC, an advisory body of the French ministry of economy) launched a working group for nanotechnology, Vivagora was one of the two civil society organizations represented in the committee, alongside a consumer group. Vivagora was also represented at the French National Agency for Research (ANR)'s committee for nanotechnology research, in the French standardization nanotechnology committee, and, later, in the nano-responsible project (see chapter 4). As a member of the European Environmental Bureau, Vivagora was also involved in the European discussions about the definition of nanomaterials.

My objective at this point is not to be exhaustive and present in detail all the projects that Vivagora was conducting. Rather, it is to point to the evolution of the activities of the organization, and its shift from its initial position of public debate organizer to that of an actor engaged in the making of both the "problems" and the "publics" of nanotechnology. This was the explicit objective of many of Vivagora's projects. The "Open Innovation" project was conducted in partnership with a cosmetics company that agreed to enter a process of "collaborative design" for one of its products. "Coexnano" was a "pluralist expertise" process, funded by the French ministry of the environment, during which Vivagora brought together representatives of environmental movements in order to interview industrialists in the construction sector about the use of nanosilver and nano titanium dioxide in paints and coatings.

In all these activities, the organization was actively engaged in the problematization of nanotechnology. It sought to make it a public issue, on which social movements could have a say, and which could lead to a transformation of the innovation processes (both at the policy and technical levels). The theoretical and practical difficulties related to this objective were reflected upon by the organization itself. During one of the 2009 general assemblies, the president of the organization, historian and philosopher of science Bernadette Bensaude-Vincent, explained, "This is a globalizing process. Everything is included: social sciences as well, as they are asked to monitor the changes and the evolutions, and tailored to the overall objectives of development and growth. Everyone becomes a stakeholder, everyone is included. There is no exteriority any more. Even for ourselves as citizens. Then the question is how can we adopt the position of the critique? I think we don't have many choices. We have to act from within, and experiment with new methods."

Affirming the "no exteriority" motto shifted the form of mobilization from the organization of public debates to participation in the construction of nanotechnology. Such a choice—opposed to that of PMO—had particular importance for Bensaude-Vincent. As a historian and philosopher of science working on nanotechnology, she wrote books and papers on the topic, and was regularly asked to talk publicly about nanotechnology. Her own engagement as president of Vivagora was never unproblematic but intersected in complex ways with her scholarly work. Listening to her, taking notes, and participating in the collective discussion about the impossibility of the exteriority position for Vivagora, I was in the same quandary. This meant that as social scientists, Bensaude-Vincent and I were always caught in the same problem of distance the other members of Vivagora faced when working with industrial or administrative actors. For the organization, "being included" meant that it had multiple links—including funding ones—with private or public institutions. It forced Vivagora to be constantly involved in negotiations about the nature of its mobilization, while under the continuing threat of being used as an alibi. For the researchers involved, "being included" was both a condition for the empirical work about the organization, and a trial of one's own engagement with nanotechnology.

For the social scientist, the practical problem of inclusion is manifest in situations where he or she is expected to "give voice" to the organization. I was indeed caught in situations in which I could speak for it, and others in which I could not. Invited by Vivagora to participate in the Nanoforum process, I could insist on the critical examination of instruments like nanoparticle labeling, as I thought it was necessary in order to critically account for the development of nanotechnology. In the somewhat informal organizing committee (in which other academics were also present and which did not have the rigid nature of a long-standing administrative body) I could negotiate the specificities of my position as both a member of Vivagora and as an academic, and feel comfortable with the research environment I was a part of. Throughout the various exchanges with the organization, my interventions contributed to the evolution of the Nanoforum, as well as Vivagora, while the various projects of the organization transformed the research I was doing. However, such relative ease to speak with and for the actors did not easily translate to other situations, in which "traditional" forms of representation were expected, as, for instance, when Vivagora was looking for someone to speak in its name during hearings conducted by a French administrative body. My engagement with the organization, provisional and explored through constant negotiations with Vivagora, was never a given. As the organization was experimenting with various types of distance with the public and private actors of nanotechnology, so did I, when accepting or refusing to represent the organization.

Accounting for his own relationships with the French Muscular Dystrophy Organization, which he studied at length with Vololona Rabeharisoa, Michel Callon speaks of the dual "engagement/detachment" strategy of the analyst, who produces social realities with the actors he studies/works with, but nonetheless needs moments of detachment in order to write accounts of the interactions, craft his own repertoire of description, and confront empirical cases with one another (Callon 1999; Rabeharisoa and Callon 2004). The case of Vivagora points to the many adjustments, the multiple negotiations and the microtrials that are part of the day-to-day interactions with the actors, and necessary conditions for the stabilization of a situation of work and analysis. For articulating attachments and detachments is clearly not easy or straightforward. My own experience with Vivagora demonstrates some of the difficulties it may entail, and that switching from "attached" to "detached" requires permanent adjustments with the actors. As much as the organization needs to constantly question its form of engagement with administrative, industrial, or civil society actors, I had to constantly question the modalities of my own engagement with Vivagora (and, consequently, nanotechnology) when studying and working with the organization.

A Temporary Intervention in the French State Experiment with Nanotechnology

By transforming its mode of social mobilization from the organization of public meetings to an active intervention in the problematization of nanotechnology, Vivagora continued the French experiments with the public management of nanotechnology within the practices of social mobilization. In doing so, it directly contributed to the extension of the state experiment with nanotechnology in France. Take, for instance, the national debate on nanotechnology. At first, Vivagora supported the process. It participated in a contribution written under the aegis of the Nanoforum and released on the CNDP website, which framed the then-upcoming public debate in terms of the exploration of the uncertainties related to nanotechnology. Vivagora then grew more critical, and its director and president eventually wrote a tribune in *Le Monde* claiming that, while CNDP was "legitimate," the participatory device itself was not adapted to large-scale issues such as nanotechnology, and proceeded to blame the government for not questioning enough the very objectives of nanotechnology development.[40] Yet the varieties of Vivagora's position made it very difficult to stabilize a consistent form of intervention. Criticized both by the radical activists (for participating in the development of nanotechnology) and by public officials (e.g., during the public debate, for not supporting the process throughout), Vivagora's experiment with social mobilization on nanotechnology required constant care, and an ability to permanently rephrase the objectives and practices of the organization. This fine-tuning eventually proved difficult to sustain for an organization in constant search of financial resources and allies in both the industrial and the civil society worlds.

Nanotechnology Trials for Social Mobilization

Nanotechnology is a trial for social movements. Environmental movements wishing to push for more stringent regulation of nanotechnology
need to make visible the risks of nano substances and products, and, consequently, the substances and products themselves. This means that any
mobilization on the environmental or health impact of nanotechnology
will necessarily result in the participation in boundary-making for the definition of the nano-ness of substances and products. This implies that these
organizations enter the sites (legal arenas, standardization institutions,
European regulatory bodies) where the definition of nanotechnology
objects are discussed, and adopt a form of mobilization that solidifies a
stake on which it can fight for (e.g., "nanosilver is a new substance," or
"300 nm is an appropriate upper size limit to define nano-ness").

Such a construction of "publics" and "problems" adopts the form of
negotiation among stakeholders about risk issues. It is part of nanotechnology as a technical domain about which regulation is discussed. As seen in
the previous chapters, it might conflict with the mobilization of the "broad
public" by science policy officials, and, consequently, with the objective of
the collective and consensual construction of nanotechnology. In Europe,
for instance, the insistence of the European Environmental Bureau for more
stringent regulation, and, in parallel, the preferred route of the "scientific
understanding of the public" on the part of the European Commission
illustrate differences in the vision and practice of the adequate representation of the European public. In the United States NGO actors explicitly
disregard the intervention of science museums (see chapter 2) or the small-
scale social science experiments meant to demonstrate the interest of real-
time technology assessment (see chapter 6). As Jaydee Hanson, in charge of
nanotechnology at the International Center for Technology Assessment,
said, "They claim they want to listen to the public, but that doesn't make
the nano people listen to what we say."[41] For Hanson, the mobilization of
ICTA through the petitions it sent to EPA was the only way for civil society
to make itself heard, that is, out of the scope of nanotechnology programs'
initiatives aimed to the construction of publics.

For all the oppositions among civil society and industrial and administrative actors, the problematization of nanotechnology that these forms of
mobilization propose is based on the discussion around the modalities of
the definitions of nanotechnology objects. As such, it directly fits within
collective discussions that occur at ISO, OECD, EPA, or the European
Commission. This does not mean that from their own viewpoints, the

conditions of NGOs' interventions could not be made easier.[42] But it does illustrate that NGOs can play the negotiating game in the definition of "nano-ness," through the defense of stakes by organized stakeholders. Hence, the nanotechnology trial can be passed, through the transformation of nanotechnology into a topic of negotiation under the adversarial format EEB or ICTA are used to.

The examples of mobilization that we encountered in Grenoble do not follow this pattern. Equalizing negotiation with integration in the making of nanotechnology, anti-nanotechnology activists consider that what matters the most is the construction of a distance from which critical inquiry can be performed. Starting from the very same refusal to reduce nanotechnology to a matter of negotiation on well-defined stakes, Vivagora contends that the exteriority position cannot be sustained, and that, consequently, the object of social mobilization is the construction of nanotechnology's problems and publics. In both cases, the position needs to be stabilized through constant adjustments. In both cases, the problematization of nanotechnology is less about negotiating about the nano-ness of substances and products than about considering—pretty much as this book does—nanotechnology as a global program of development gathering objects, futures, concerns, and publics. PMO considers that the global character of nanotechnology requires putting all its components (including participatory mechanisms and stakeholder negotiation processes) at a distance. It criticizes the agencements described throughout the book (be they participatory instruments or risk management methodologies) for the connections they perform between the development of nanotechnology and the engagement of citizens, between science and science policy programs, between science and social science. In turn, PMO hopes to propose a "pure" critique at a distance, which would avoid the complex arrangements nanotechnology is made of, but which requires constant care to be sustained. The agencement on which it relies is never a given, as PMO encounters participatory devices that aim to integrate even the most critical positions. Vivagora, on the other hand, considers that social mobilization has to cope with the impossibility of being exterior to nanotechnology. This requires experiments to produce publics and problems, intervention in the very making of agencements, and, consequently, results in permanent uncertainty about the identity and objectives of the organization. Whereas PMO works hard to distance itself from any form of organized social mobilization, Vivagora hopes to "structure civil society" by circulating information, meeting with representatives of environmental social movements, participating in collective actions in the United States and Europe, and inviting members of NGOs to speak at events it organizes or to participate in projects it undertakes. The mobilizations of PMO and Vivagora problematize nanotechnology through practices of engagement and modes of collective and individual actions. How to mobilize is then part of what to mobilize on. The spectacular demonstrations and critical inquiry are part and parcel of the critique of the global program of nanotechnology, as much as the various experiments undertaken by Vivagora are components of its mobilization on the "democratization of nanotechnology." Problematizing nanotechnology through social mobilization is at the same time problematizing social mobilization itself. Eventually, the forms of social mobilization encountered in this chapter participate in the making of political actors acting within the French state experiment on nanotechnology. Whether they adopt a Rousseau-ist political philosophy and thereby refuse the very terms of this experiment, or attempt to engage in it by rethinking the role of civil society, the organizations we encounter are part of the institutionalized practices of the French democratic life.

In such a process, the engagement of the actors crosses that of the social scientist. Whether he or she adopts a position at a distance to describe activists or engage with others in the experimentation of agencements, the problematization of nanotechnology and social engagement in it is, by the same token, that of the engagement of the analyst. These two organizations are particularly interesting as both of them, for all their differences, ask similar questions as those of this book: how to describe nanotechnology? How to understand its relationships with the construction of democratic order? How to envision a path toward the democratization of nanotechnology, and technology more generally?

How to learn from the experiences described in this chapter for our own development of the democratic analysis of technology and our form of engagement and critique? This will be examined in the next chapter.

7 Democratizing Nanotechnology?

Sites of Problematization

In chapter 6, we encountered various initiatives meant to sustain social mobilization within or against nanotechnology. These initiatives directly echo the main concerns of this book: how to critically describe large-scale technological programs associating objects, futures, concerns, and publics? How to identify the questions they raise for democracy and envision perspectives for democratization? These questions articulate the practice of social scientific research with the issues of research ethics and critical engagement. This chapter builds on the empirical inquiries of the previous chapters in order to question what the democratization of nanotechnology could and should mean. It connects the sites examined in the previous chapters in order to identify contrasting problematizations of nanotechnology and democracy. While these problematizations do not claim to be exhaustive, they do offer insights into current evolutions of contemporary democracies. And, perhaps more importantly, they will help me reflect on a nonevaluative yet normative perspective that I eventually characterize as a critical constitutionalism.

In the previous chapters, I have conducted the analysis in the sites where nanotechnology was problematized. These sites of problematization were science policy offices, science exhibits, participatory mechanisms, standardization and regulatory bodies, expert organizations in science or social science, and places of social mobilization. They are the places where nanotechnology is defined as a problem deserving a range of acceptable solutions, and consequently, where democracy itself is problematized. This has led us to explore places not always associated with democratic practices (such as standardization organizations and science museums) or usually situated at its fringes (such as public expert bodies). But it is precisely in these margins that the rules governing institutions such as the European

Commission, national governments, and international organizations have to be questioned when dealing with nanotechnology's objects, futures, concerns, and publics. It is in these sites that questions of citizenship, legitimacy, and national or European sovereignty were explicitly raised. These sites are not passive scenes, on which problematization of nanotechnology would be stabilized or destabilized. Consider, for instance, the public meetings of the French national debate on nanotechnology. The separation that the organizers maintained between the invited speakers and the public in two different rooms and the eventual closure of the debate were part of the problematization of nanotechnology they propose. Hence, sites are not a priori distinct from the agencements that problematize, and their natures and rules can be contested. They are part and parcel of the processes of problematization. Their variety displays the extent of places where democracy is at stake in contemporary societies. How to make these sites the location of a critical reflection on democratization?

Throughout the sites I examined in the previous chapters, we encountered agencements that define the "nano-ness" of objects, produce anticipations about the future, identify public concerns, and shape the agency of the political subject. Speaking of agencements was an approach to studying the problematizations of nanotechnology through the instruments that problematize, in ways that did not separate "reality" from its "representations." For instance, when I explored diverse definitions of "nano-ness" in chapter 4, I described various ways of granting regulatory existences to nanomaterials. Analyzing the replication of technologies of democracy in chapter 3 was an investigation into the making of political subjects, whether "deliberating" or "debating" citizens. By highlighting processes of problematization, the analysis was inserted in the making of the objects, futures, concerns, and publics of nanotechnology.

STS thinkers such as Annemarie Mol or John Law would speak of the "ontological politics" at stake here, in that "reality does not precede the mundane practices in which we interact with it but is rather shaped within these practices" (Mol 1999, 75). The expression points to the contingency of choices related to the shaping of objects and subjects, and concurrently, to the possibility of conceiving these realities differently. It is an invitation to consider the forms of scholarly engagement as "interferences," an expression used by John Law to point to the many connections between the description work and the activities of the actors themselves, and the performativity of social science (Law 2010, 278–279). For Law, focusing on "interferences" directs the attention to the contingency and particularities of the situations at stake. It prevents one from taking for granted the dichotomy between "description" and "intervention." One can indeed understand as "interferences" the analysis of nanotechnology that I conducted and presented in the previous chapters, intervening in sites of problematization alongside the actors involved (as in the OECD or in Vivagora's public initiatives) or contributing to their publicity by describing them. Thinking in terms of interferences pays close attention to the local situations of uncertainty for the analyst's engagement. It adds another dimension to the ontological politics at stake within sites of problematization, namely that of the politics of scholarly engagement.

The next research steps could then consist in delving further into the exploration of the multiplicity of problematizations. There could be two ways of doing that. One would be to concentrate on some of the sites I studied, and describe at further length the microprocesses that led to the expression of the variety of problematizations in each of the sites. A second would be to multiply the forms of interferences.[1] Choosing one approach or the other, the natural conclusion could well be that "reality is multiple," that various problematizations of nanotechnology are proposed, and that, consequently, "decisions" and "choices" are all situated, and distributed in heterogeneous processes. This is not a position that I find satisfactory, for both analytical and political reasons.[2] What would be gained in terms of the quality of fine-grained descriptions would prevent an analysis of nanotechnology and democracy as categories extending over wide institutional spaces. It would limit ontological politics to the local sites of trials, while there is indeed ontological politics at stake in the making/stabilization of such entities as "states" or "international organizations." It would prevent such crucial questions as: how do democratic experiments acquire value? For whom are they valuable?

Answering such questions requires an examination of the spaces within which such initiatives as the French national public debate on nanotechnology, the nanotechnology exhibits in American and European museums, the regulatory attempts at governing nanomaterials or mechanisms based on "real-time technology assessment" are valued and for and by whom: who are the audiences in front of whom these initiatives are conducted? In whose name and for whose interests? What public spaces are then crafted?[3] These questions are directly dealt with in the sites of problematization of nanotechnology encountered in the previous chapters. By considering that reconstruction is part of both the analytical and engagement work of the social scientist, as he or she circulates across sites and draws connections among them, this chapter builds on the empirical explorations of sites in

order to reconstruct spaces characterized by common problematizations of
both nanotechnology and democracy.

Four Problematizations of Nanotechnology

Mobilizing Expertise to Realize the Potential of American Nanotechnology

When the National Nanotechnology Initiative (NNI) was created in the
late 1990s, its integration in the making of American science was manifest.
Nanotechnology was "the next frontier," as announced as early as 1959 by
physicist and Nobel Prize winner Richard Feynman, the "next industrial
revolution," for which society had to be prepared. The value of the new
technology was to be demonstrated in order for members of the U.S. Congress to fund the initiative, for students, workers, and consumers to participate in its development, and for the "general public" to accept it. One
could identify the connection with past science policy programs in the
United States, from the Apollo project to Vannevar Bush's vision of science
as an "endless frontier."[4] But nanotechnology is a particular case. It is not
meant to be a government-driven program aimed toward the realization of
a single objective (like the Apollo or the Human Genome Project). Nor
does it follow a linear, science-based model that would contend that funding basic science is a sufficient condition for the development of applied
research, and, eventually, social progress. The NNI is best described as a
program that operationalizes in research management instruments long-
term objectives, research organization plans, and understandings of the
historical development of science and technology. It associates numerous
federal agencies, and brings together fundamental and applied science for
the development of nanotechnology. Nanotechnology objects and futures
have caused vivid controversies between industrialists and proponents of
visions of nanotechnology based on the anticipation of self-replicating
molecular machines (see chapter 1). Eventually, the instruments of the
NNI were able to connect both, while making nanotechnology a vast
program gathering a large number of projects. For instance, Roco's four
generations of nanomaterials connected current practices, industrial applications, and long-term developments. It allowed the NNI to avoid the
long-term and scary visions of Drexler while also situating nanotechnology in the continuation of Feynman's prophecies, within the history of
scientific discoveries.

Yet nanotechnology also caused public actors to deal with concerns. The
proponents of nanotechnology in the American science policy landscape

soon advocated the management of risks and ethical issues through specific expert work, and the integration of "public input" in nanotechnology programs. As the example of nanosilver illustrates (see chapter 4), the way of dealing with these issues is, in many respects, defined within the American expertise system in federal agencies. Legal conflicts occur on the qualification of substances (as "new pesticide" or "known material" in the case of nanosilver), and the legal arena is the terrain on which arguments are presented and opposed to each other, and administrative choices are challenged.

Nanotechnology is problematized, in the American sites we encountered, as an issue of scientific development for the sake of collective progress, and for which the American society has to be prepared, and externalities are to be taken care of. Doing so implies the use of appropriate expertise, and this requires demonstrating the quality of the expertise being mobilized and the scientific value of the stakeholders' positions. This process was described in chapter 4 about silver nanoparticles, and in chapter 6 about expertise in ethics. It enacts boundaries between expertise and the stakeholders' interests, and between the expertise in ethics and toxicology and the technical development of substances. It makes the relevance of the expertise to mobilize a matter of public discussion. Hence, numerous organizations called for the integration of more federal funding for environmental, health, and safety (EHS) research during the discussions that led to the reauthorization of the National Nanotechnology Initiative by Congress in 2009.[5] But mobilizing the "good expertise" to answer nanotechnology concerns is not only a problem of research funding. It also requires that the expertise be identified. Thus, successive congressional reports interrogated the quality of the reporting of EHS activities in nanotechnology programs, and, by 2008, called for a better monitoring, identification, and quantification of EHS research.[6] A specific area of expertise was needed, and it was to be visible enough for policymakers to mobilize, evaluate, and control it.

Through her exploration of biotechnology policies, Sheila Jasanoff has identified the components of an American contentious civic epistemology (Jasanoff 2005). She demonstrates that the processes through which public decisions gain scientific objectivity and democratic legitimacy are based on a combination of an adversarial style of policymaking, mobilization of expertise as a way of escaping politics, and calls for transparent decision-making processes. One can see these dynamics at play within the problematization of nanotechnology in the United States, which thereby appears as an empirical lens for scholars interested in the description of the American civic epistemology. Yet the problem of identifying nanotechnology objects and dealing with strong connections to future prospects introduced some displacements in the known processes of objectivity and legitimacy building. We encountered several of them, when proponents of nanotechnology suggested a "safety by design" approach to considering toxicological properties at the design phase of substances; at the Boston Museum of Science, where science communication specialists make "deliberation" an important part of the public communication of nanotechnology; and at Arizona State University's Center for Nanotechnology in Society, where "real-time technology assessment" is conceived as an intervention in the construction of science and society meant to "democratize science" (Guston 2004). The last two experiences explicitly envision democratization as their objective, while the first is conceived as an innovative way of anticipating social concerns from the core of technological research.

These attempts could question the boundaries between expertise and politics, and indeed the very definition of the relevant expertise to mobilize on technological issues. Their impacts are not given, and require adaptations so that innovative interventions can be integrated in the space of expert knowledge production, and, eventually, democratization be reached through the development of additional technical competencies in the public management of technology. Making safety by design a new area of expertise for dealing with nanotechnology risks would require scientometrics methods able to render measurable safety-by-design projects, which, by definition, are not understood by the distinction between material sciences and toxicology.[7] The uncertainty surrounding the objectives of deliberation as introduced in the American science museums is dealt with through the transformation of deliberation into an expertise managed by museum staff, and addressed to individual citizens expected to learn about a new scientific field. The construction of a small-scale experiment at the Center for Nanotechnology in Society (CNS) described in chapter 5 can be seen as a demonstration of the scientific quality of an approach that does not separate science from society, but can nonetheless differentiate its expert work from public decision making.

A French State Experiment with Nanotechnology
As French public institutions attempted to make nanotechnology both a program of technological development and a governable domain, they also made it a problem of engaging the French democracy, at local, national and European levels. The importance of centralized technology policy and research initiatives for the development of the French state has been

described in other domains than nanotechnology (see Hecht 1998). But the situation is complex for the centralized expertise of the French state, as the interests of the concerned publics are not well identified and the objects at stake not defined, while radical activists question the role of public bodies. At local levels, the involvement of local public bodies in nanotechnology has to cope with protests, and the science communication specialists, as in the Grenoble science center, are at pains to transform the relationships with science and its publics. How to deal with the social and technical uncertainties of nanotechnology? How to define a national position in European and international arenas? These are the questions that the French state is expected to answer.

In chapter 3, I described the national public debate on nanotechnology organized in 2009 as a state experiment. The replication of the CNDP public debate procedure on nanotechnology engages the modes of intervention of the French state and indeed its very nature. Similarly, the introduction of the *substances à l'état nanoparticulaire* ("substances in a nanoparticulate state") category, the nano-responsible project, and the science exhibits turning visitors into debating citizens transform the roles of public bodies and are signs of the attempted extension of their competences to new areas—poorly defined chemicals, uncertain industrial processes, and unknown publics. As it struggles with nanotechnology's objects, futures, concerns, and publics, the French state experiments with the ways and means of its public action. In replicating technologies of democracy or introducing innovative techno-legal instruments, the French state is also experimented with. This state experiment manifests itself in the integration of new components in French state expertise, with the objective of governing social and technical uncertainties, be they unknown concerned publics or uncertainly defined *substances à l'état nanoparticulaire*.

This makes nanotechnology a component of a wider evolution. Thus, the director of ANSES, the public health agency in charge of the mandatory declaration of substances in a nanoparticulate state wrote in the national newspaper *Le Monde* that "experts are not researcher monks" (*moine chercheurs*). He meant that expertise could not remain in the secluded place of research, but needed to answer social problems, be aware of controversial situations, and make sure that the involved actors are heard. He argued, "this is precisely by enlarging the space of controversy, as a place for well-argued discussion, that we will avoid polemic."[8] The opposition he drew between "controversy" and "polemic" is significant. While the latter was characterized by irrational exchanges of opinion, the former could be organized as a collective process of political and technical rationality. This was, for him, precisely the role of the public agency—illustrated by its intervention in the field of nanotechnology, through its role in the development and management of the *substances à l'état nanoparticulaire* initiative.

This proposition has a particular resonance in France, where centralized public expertise is a basis of the democratic state.[9] It is inscribed in, as much as it contributes to shape the trajectory of a powerful state, expected to guarantee the neutrality of administrative expertise, and prone to integrate new concerns in this very expertise. Political scientists have described how the French state managed to integrate environmental issues related to industrial activities into the centralized public administration of industry (Lascoumes 1994). Others have analyzed the response to health crisis in the 1990s and showed that the French state created health agencies meant to ensure the neutrality of its technical expertise while also taking demands for a greater public participation into account (Benamouzig and Besançon 2003). The creation of CNDP in 1995 and the extension of its missions to general policy options in 2002 are steps in the development of "the French experiment with public participation," by which the state relies on expertise about participatory matters (Revel et al. 2007). These evolutions display a state constantly attempting to integrate new components in a centralized expertise that grounds the legitimacy of its intervention. This powerful state is able to act through an expertise owned by various government components, public agencies, and research organizations, brought together for the sake of the development and control of technology. It attempts to govern technical objects uncertainly defined, and to constitute new political subjects by turning poorly identified publics of science into debating and participating citizens.

The outcome of the French state experiment with nanotechnology is still uncertain, and what will appear out of it unsure. This is particularly clear when considering that the civil servants involved in nanotechnology-related issues are permanently raising questions about their positions, and the objectives and modalities of public policy actions regarding nanotechnology. They gather in informal working groups across ministries, participate in public meetings, intervene in European and international arenas where they represent France and argue for specific "nano" regulation in Europe and (unsuccessfully) for international initiatives able to take the technical and social uncertainties of nanotechnology into account. Over the past few years, many of the French civil servants involved in nanotechnology I interviewed were keen on making this domain a new area of intervention for the French state. But they were also anxious how so much uncertainty could be made governable and were wondering about the overall perspective of their works. The uncertainty about the outcome of the state experiment with nanotechnology also impacts the intervention of civil society organizations such as Vivagora (see chapter 6). While the organization intervened in various ways in the democratization of nanotechnology, the absence of a known and consensual procedure for collective discussions about technological development was both a condition for its experimental interventions and a source of permanent uncertainty about the roles of social mobilization.

That the outcome of the state experiment with nanotechnology is uncertain is a sign of the incomplete transition of the French state, which is imperfectly equipped to deal with the new entities it attempts to make governable. It makes it easy for proponents of nanotechnology to ignore the attempts at governing social and technical uncertainties. Consider, for instance, an initiative undertaken by CEA, unironically called "Nanosmile." Developed as part of a European project, Nanosmile was an online training device meant to describe the approach to be taken in order to "apprehend potential risks and benefits of nanomaterials in order to contribute to Science & Society dialogue" (Laurent 2010, 85). It separated the "subjective perceptions" from the "objective risks" to be mastered by "good practices." For the proponents of Nanosmile, what mattered was the production of adequate representations of science, for the benefits of known publics, namely ignorant crowds prone to irrational concerns. In Nanosmile, social and technical uncertainties were not in the picture.

The incomplete evolution of the French state makes it particularly vulnerable to criticisms, and the anti-technology activists were particularly vocal about nanotechnology. For them also, uncertainty is not an issue. They consider that the attempts at governing uncertain chemicals and unruly publics are only signs of the blurring of boundaries among scientific research, state intervention, and citizen involvement. For them, the rationality of the French citizen is situated outside of technology development, as he or she ought to perform critical inquiry from a distant position. Their interventions are forceful reminders that a stream of political philosophy, whereby the equality of simple citizens accepting the primacy of the general interest is the basis of social order, may be threatened by the current transformations of the French state.

A Problem of European Integration

Speaking of the harmonization project as crafted in the early 1990s by the European Commission, Andrew Barry describes it "both as a way of imagining and of reordering European space, as well as a technical process directed at establishing this space as a governable entity" (Barry 1993, 316). Harmonization points to a set of operations meant to ensure the integration of the European political, economic, and moral space. It is based on such instruments as the standardization of products circulating on the European market, the coordination of policy choices of member states, or the identification of common values for European societies, such as "sustainability" or "competitiveness." We encountered some of these instruments in the previous chapters, as European public bodies proposed definitions for nanomaterials, introduced codes of conduct for nanotechnology research, or set up "networks of excellence" to coordinate the initiatives of member states. In the sites where these instruments are crafted, the problem of nanotechnology is that of the composition of the European harmonized space. In that sense, it is a problem of integration. This integration is political, economic, and moral. It is political in that it relates to the way European institutions can define long-term objectives (such as the development of new technological domains or the promotion of values deemed European) and exercise control over scientific and technical activities that, according to the subsidiarity principle, are governed at member state level. Integration is economic in that technological development is expected to make Europe a place where laboratory research is transmuted into market developments and technological innovation meets the needs and expectations of the European consumer. Eventually, integration is also moral in that European values are expected to be integrated at the core of technological research. Defining these values also defines what it means to qualify objects, people, and practices as "European."

Problematizing nanotechnology as a matter of integration makes nanotechnology a step in the transformation of the European research policy. Commenting on the report about converging technologies he edited, philosopher Alfred Nordmann spoke of a "European experiment" (see chapter 5). The expression is accurate, as it points to instrumented interventions in the making of Europe itself, for still-uncertain results. These interventions are much wider than Nordmann's report, or indeed nanotechnology. There are situated within long-term reflections pertaining to the nature of Science–Society relationships in Europe (see Felt and Wynne 2007), and more generally to the appropriate way of making science and technology engines of European integration (Laurent 2016a).

As seen in chapter 5, nanotechnology paved the way for the development of "responsible research and innovation" (RRI), itself a component of the post-Lisbon strategy recompositions of the European research policy. The 2000 Lisbon strategy, which hoped to make Europe a knowledge-based economy, defined target levels of public and private investment in R&D (3 percent of GDP) for member states. Evaluated in the early 2010s, it was considered a failure since the majority of member states never reached this target share. By contrast, the new forms of the European research policy after 2010, notably within the Europe 2020 strategy, did not attempt to define minimal thresholds of investment, but instead target a limited number of objectives considered to be priorities (Lundvall and Lorenz 2011). This evolution had a dual objective, which directly resonates with the problematization of nanotechnology described in the previous chapters.

First, it is situated within the same line of argument as the RRI: European science has to adapt to what the European public could consider meaningful. Targeting "challenges" such as global warming or aging is a way of doing so. In chapter 2, I described the "scientific understanding of the public" that the Directorate-General for Research and Innovation of the European Commission defined as an objective in the wake of initiatives in nanotechnology and public communications. Knowing European publics "scientifically" was an answer to a perceived trust issue—a concern that is regularly expressed in European policy documents. Thus, the presentation of the Europe 2020 strategy relates the definitions of "challenges" supposedly meaningful to the European public to the fact that "the percentage of European citizens [who] trust science and technology to improve their quality of life decreased over the last five years from 78 percent to 66 percent." The report thus considered that there was "a genuine expectation for science to reorient its efforts to contribute to addressing the societal challenges of our time."[10]

Second, the reorientation of the European research policy also offered a way of dealing with a constrained budgetary situation. Thus, Maire Geoghegan-Quinn, commissioner in charge of science and technology, said in 2011 that "at a time when most Member States are confronted with strong budgetary constraints," it was necessary to target public investments toward "growth-enhancing policies that get excellent value from the money invested, prioritizing the most cost-effective reforms that help develop new markets for innovative products and services." In this declaration, the evolution of the relationships between European science and the European publics appeared as a part of a broader recomposition of the European economic policy.

Thus, the European sites of problematization of nanotechnology are steps toward the redefinition of the European research policy, and more generally, participants in the construction of Europe as a consistent space. They make the integration objective less a matter of uniformity of levels of research investment across member states (as in the Lisbon strategy) than a problem of political, economic, and moral harmonization across the European Union. Integration is expected to answer many concerns, whether they are related to the much-discussed European democratic deficit, to the EU's economic strength, or its problematic common identity. It is, then, the vehicle for making Europe a common space, within which member states, private companies, and European publics are closer with each other and with the European institutions.

Realizing integration relies on a mode of reasoning whereby the European public action needs to ensure a balance between constraining legal interventions and delegations to market mechanisms. Coordination devices such as codes of conduct and ethics review (see chapter 5), and the regulatory precaution approach in the treatment of nanomaterials (see chapter 4) are instruments expected to ensure this balance. They can be described using the works of political scientists who characterize a European style of policymaking as "experimental governance." Commenting on instruments such as the Open Method of Coordination, through which the European institutions determine broad policy objectives and implement a set of instruments (guidelines, benchmarks, etc.) for member states to reach these objectives on a voluntary basis, these scholars argue that experimental governance aims not to legally constraint but to coordinate actions in reversible ways.[11] The experimental governance literature often considers that experimentalism can unproblematically be equated with greater efficiency of policymaking, and better democratic practices. Yet the European experimentalism encountered in this book is also a terrain of oppositions, about how to craft interventions that would achieve an appropriate balance between constraining legal actions and the delegation of collective organizing power to markets. One can read the opposition between the European Commission and the European Parliament in those terms (see chapter 4). The explicit analogy between "scientific understanding of the public" and market studies is at odds with more sophisticated approaches that attempt, for instance, to craft a "lay ethics" that would leave social expectations open (Davies, Macnaghten, and Kearnes 2009; see chapter 5). Therefore, if nanotechnology can indeed be considered a "European experiment," it is not because Europe would be the place where science and society would finally come together and produce, at last, a democratic technology policy. Nanotechnology is a European experiment in that it entails explorations of the channels of democratic legitimacy in Europe, and, more generally, of defining and governing Europe itself.

International Nanotechnology for a Global Market

In June 2009 in Braga, Portugal, the OECD Working Party on Nanotechnology (WPN) organized a roundtable on international cooperation in nanotechnology.[12] The chair of WPN at that time was physicist Robert Rudnitsky, who was also chairing the Global Issues in Nanotechnology working group at the NNI. Rudnitsky gave the opening talk, in which he equated international cooperation primarily with an operation protecting the development of nanotechnology from foreign threats. "Previous technologies have seen public acceptance of rejection begin in one country and migrate to others," he said, and "international regulatory regimes affect U.S. industry." Developing "a healthy global marketplace for U.S. nanotechnology goods and products" required international cooperation in order to avoid these troubles. The example he had in mind then was that of biotechnology, and particularly the controversies about GMOs. In the reading of these episodes by policymakers involved in nanotechnology programs, the biotechnology experience is that of a failed harmonization. Opposition over GMOs in Europe resulted in differences in regulatory choices, people like Rudnitsky claimed, which hindered market developments and resulted in disputes in front of the WTO.

This understanding of the biotechnology case can be criticized. It ignores the embeddedness of regulatory choices and public reception of technology in stabilized institutional constructs (Jasanoff 2005), the variations in the construction of science as a basis for decision making (Winickoff and al. 2005), and the subtlety of the anti-GMO critiques on both sides of the Atlantic (Marris 2001; Joly and Marris 2001). Consequently, it oversimplifies the trajectory of technology acceptance and tends to make it an issue of public fad followed by irrational rejection (Rip 2006). But however inaccurate this narrative of the GMO case might be, it is crucial in the problematization of nanotechnology as an issue of international cooperation. It is used as a counterexample demonstrating the need to anticipate potential threats to the extension of a global market, be they differences in regulatory choices or variations in public acceptance—both potential sources of trade barriers.

As Rudnitsky was speaking in Braga, he identified an objective of "harmonized policies and constructive interactions between nations." But "harmonization" was different in his speech than in the European case, where it is an operation expected to craft a common European identity. Here, it refers to the objective of "developing an international marketplace for nanotechnology products and ideas." The 2011 U.S. strategic plan for nanotechnology similarly referred to the need to "increase international

engagement to facilitate the responsible and sustainable commercializa-
tion, technology transfer, innovation, and trade related to nanotechnology-enabled products and processes" (National Science and Technology
Council 2011, 27).

In this document, international collaboration was seen as a condition
for the "development of a vibrant and safe global marketplace for nanomaterials and nanotechnology-enabled products." International cooperation
for the construction of a global market requires common terminologies and
standards, which are crafted at the International Standardization Organization (see chapter 4). International cooperation also has to be conducted at
policy levels. Within the global objective of the development of a market
for nanotechnology, international cooperation about public engagement
can be seen as a way of preventing differences in public acceptance (see
chapter 3).

Matthew Kearnes and Arie Rip cited as an aspect of the responsible
development discourse "the way it operates internationally as a tool for the
development of global consensus and strategy" (Kearnes and Rip 2009).
The objective of "responsible development" is indeed shared, and serves
as a common reference. Yet as the American, French, and European cases
show, "responsibility" may point to different problematizations of nanotechnology. And as the intervention of Rudnitsky at the OECD made clear,
national interests are central in the development of international markets.
Thus, "international cooperation" in the responsible development of nanotechnology is not just a matter of peaceful agreements among countries
interested in the safety and acceptability of technological innovation. It is
also a strategic matter of governments and private companies eager to
ensure their market share in the developing nanotechnology market.

In France and the United States, as well as within the European institutions, problematizing nanotechnology is also problematizing democracy.
We saw that these problematizations engage the forms and conduct of
American public expertise, the nature of the French state, and the identity
of a European Union in the making. That nanotechnology is a matter of
market making at the international level does not mean that democracy
is not at stake. First, international organizations develop reflections and
expertise about democracy. At the OECD, the WPN makes "public engagement in nanotechnology" a matter of democratic practices. It did so by
crafting an expertise on technologies of democracy, separated from the
content of nanotechnology issues. A "policy expertise" could then be proposed in ways that would not cross the boundary between the international work and national policy choices. Second, the projects conducted

within international organizations are expected to be based on collective negotiations, in which public and private interests are represented. This results in a particular political format, which relies on science to produce international consensus. The "science" on which the international expertise and standards are supposed to be based is made of heterogeneous considerations. Recall, for instance, the mixture of science policy logic, communication imperatives, and technical considerations that had to be mobilized so that the 100 nm size limit for the definition of nanomaterials could hold at ISO TC229 (chapter 4). Eventually, it prevented an association between the fact of being "nano" and the eventuality of increased risks.

The particularities of the international problematization of nanotechnology appear clearly when one considers the purification devices it requires in order to eliminate propositions that do not fit with it. Thus, attempts at defining nanomaterials using properties related to size rather than size itself were not acceptable within ISO. Initiatives that would have connected public engagement with public intervention in the government of nanotechnology objects could not succeed at the OECD WPN. Eventually, the international consensus on the construction of a global market for nanotechnology requires a constant purification of international interventions expected to be untainted with policy choices reserved for the sovereign decisions of participating countries. This is not a neutral process, as it makes it impossible to define nano-ness in ways that could lead to the regulation of potential hazards, and conceives "public engagement" as no more than exercises with no effect on technological development.

Identifying Problematizations

Starting the analysis from agencements located in sites of problematization was a way of accounting for spaces characterized by common problematizations. The four problematizations of nanotechnology described previously extend over spaces that reproduce national territories, or transnational political organizations. As a global program associating objects, futures, concerns, and publics, nanotechnology is a lens for the study of what these democratic spaces are. Thus, it offered empirical entry points in the study of contemporary democracies. We encountered the importance of technological progress as a collective project in U.S. democracy—a collective project relying on trained individuals expected to become consumers, supporters of policy choices, students, or workers, and on public regulation through adversarial procedures. I described the concerns for the rationality of the French citizen, and the transformation of the powerful French state as it attempts to deal with issues of technical and social uncertainties. The description of nanotechnology as a European experiment helped characterize Europe as a political entity in the making whose democratic legitimacy and indeed its very identity are permanently questioned. Later I analyzed international organizations as sites of reflection about democratic practices and collective negotiations for the sake of the global market, where it is crucial to ground expert interventions outside of sovereign policy choices.

The problems that are discussed in these sites are different. They concern the transformation of technology assessment in the United States, the extension of state action to new social and technical entities in France, the growing European integration, and improved adjustment of a global market offer to public demands. These differences relate to variations in modalities of democratic ordering, and directly impact the construction of objects, futures, concerns, and publics that constitute technological development programs.

Identifying these differences requires two joint movements. First, one needs to avoid the apparent dichotomy between the localization of the empirical site and the macroscopic scale of democratic spaces. The previous chapters have shown that the study of sites of problematization make visible the conditions under which particular democratic experiments are valued, and the ways in which they matter for the actors involved. For instance, one cannot understand the state experiment that is the French national debate on nanotechnology (see chapter 3) without considering the redefinition of the technocratic expertise on both public participation and risk management. The identification of problematization requires that one is sufficiently close to the empirical phenomena being described in order to account for the ways local initiatives matter and to whom, and, thereby, how they participate in the making and remaking of wider spaces. Second, one needs to use the circulations across sites undertaken by actors and analysts alike in order to draw what Michel Callon has called a "political geography of sites" (Callon 2012, 151). As policymakers circulate from their national offices to international arenas, social scientists and science communication experts meet in academic and professional conferences, and market products flow across political boundaries—so the analyst needs to connect sites belonging to regulatory organizations, science museums, and public debates in order to make problematizations visible.

Stabilizing Problematizations

Foucault considered one of the main benefits of the notion of problematization to be the attention it draws toward the public restabilization processes through which problems are constantly made explicit, and solutions crafted. Within these very processes lies the possibility for displacements. In studying, in the previous chapters, the agencements that problematize nanotechnology, I examined the processes through which problematizations are constantly restabilized, and consistent spaces characterized by common problematizations are built.

One can identify three of these processes. A first one is the *extension* of known agencements. Thus, I showed how the American adversarial regulatory system included public debates about the novelty of nanoparticles such as nanosilver. I described the extension of state expertise to new domains in France, through the replication of technologies of democracy such as the CNDP procedure or the introduction of new regulatory categories. I showed how in Europe, nanotechnology is an opportunity to extend the reflection about European values to the entire research policy.

A second stabilization process is based on the *purification* of agencements. Thus, I analyzed the elimination of alternatives, be they related to definitions of nanomaterials or understanding of the objectives and formats of public engagement, at ISO and the OECD. The experts mastering technologies of democracy encountered in chapter 3 have to purify the issues expected to be discussed and the participating publics in order to eliminate anonymous activists or non-neutral panel members of consensus conferences. When an initiative such as the French nano-responsible standard enters the space of European standardization (see chapter 4), it needs to be transformed so that it becomes acceptable as a proposition based on risk–benefit evaluation.

Eventually, one can identify in *comparison* a third process of stabilization, particularly visible as European actors defined their perspectives on nanotechnology and converging technologies in opposition to the American programs. But comparison is also at stake when French civil servants compare their initiatives with those of other European member states more reluctant to act to regulate the uncertain risks of nanomaterials, or when Grenoble-based museum experts judged European projects disappointing because they do not make nanotechnology a topic of experiment in the display and practice of public debate, as French science museums attempt to do. Comparison, eventually, is constantly undertaken in international organizations such as the OECD, where national initiatives are benchmarked against one another, in the hope of constructing a global expertise.

Extension, purification, and comparison are not mutually exclusive. For instance, I describe in chapter 2 how a network of American museums extended its expertise on informal science education to new topics and methods. The comparison with the European science museums was used to call for new methods based on "two-way communication." In the meantime, the network purified innovative deliberation exercises such as the forum developed at the Boston Museum of Science in order to include them within its expertise distributed across the country. The three operations certainly do not cover the entire range of processes that stabilize problematizations. But they help point to the importance of accounting for problematizations not as given entities, but as outcomes of stabilization processes. They direct the attention to the contingency of dominant formations, as well as to the struggle and tensions among them. They also make potential alternatives and variations visible. Thus, the extension of the French state expertise to participatory mechanisms and uncertain risks is strongly resisted by anti-technology activists, who use the attempts at extension as opportunities for proposing to turn the French public into "simple citizens" engaged in critical inquiry. The purification processes needed to write international standards make eliminated choices visible for the analyst, who can then locate, for instance, property-based definitions of nanomaterials as alternatives to simpler, size-based definitions. Eventually, the comparisons undertaken in international arenas also make it possible to envision the broader range of problematizations of nanotechnology and democracy, such as the British upstream engagement discussed at the OECD WPN (see chapter 3), which would base the legitimacy of collective decisions on a public intervention at an early stage of a technological development understood in a linear way.

Extension, purification, and comparison locate the analysis in the midst of the operations that problematize, and directly raise the questions of counter-problematizations. How then to envision the critique of democracy, or, symmetrically, the possibility of political intervention in the democratization of nanotechnology in particular, and technology in general?

What Critique?

Democratic Ideals?

After having identified various problematizations of nanotechnology, various democratic constructs, and located the sites where counter-problematizations are proposed, how should one envision the critical strength of the analysis?

Claiming that the analysis can merely stand outside of the described realities, possibly for others to take sides, is not satisfactory. For the conduct of research, and indeed the very nature of sites of problematization, makes such distinction between description and intervention not relevant. Another option would be to evaluate the democratic construct according to external criteria. This would replicate the position of political theorists developing a critique of "actually existing democracy" (as philosopher Nancy Fraser [1990] would say), or an evaluation of participatory procedure according to known criteria of "good deliberation." Scholars have proposed to grade the "social robustness" of governance and participatory initiatives about nanotechnology. They consider that participatory initiatives in nanotechnology "only partially meet aspects of social robustness, and that the governance and deliberative turn in science and technology policy has not led, so far, to greater democracy and responsibility in nanoscience and nanotechnology development."[13]

Yet criteria such as "deliberation" or "robustness" are explicit parts of the problematization of nanotechnology. They are advocated by social scientists invited to give their opinions, and inscribed in controversial and diverse agencements. They enact different democratic constructions. Being "not responsible enough" is thus a weak critique, since it is entirely part of the problematization one would want to critique.

"Merely describing" problematizations or evaluating them according to known criteria are two operations based on an understanding of scholarly work and political engagement that distinguishes epistemological from normative tasks. This distinction is precisely what the analysis of problematization seeks to avoid—and had to do, given the variations of scholarly engagement in sites of problematization. This means that any critical perspective able to question the world as it is and as it should be needs to associate both the practice of research (examining questions such as: how to select sites of problematization? How to circulate across them?) and the mode of political engagement (providing guidance into the ways of democratizing nanotechnology, and indeed any other entity subjected to this kind of analysis). STS scholars and political theorists have proposed approaches that may contribute to our interrogation here. I will discuss two of them before turning to "critical constitutionalism" that, as I argue, the analysis of the problematization of nanotechnology suggests adopting.

Novelty

Describing the joint problematizations of nanotechnology and democracy, one could identify a special interest in situations where new entities are being constituted. This is a distinctive trope in STS, as some of the original works of the discipline originated from the sociology of technological innovation (Callon 1986; Latour 1992). These works considered innovation as a particular interesting domain of scholarly investigation and political intervention, since it potentially redefines the ontological quality of the human and nonhuman entities composing a given situation. In doing so, they also identified a political value in innovation, as the provider of situations where democracy might be reinvented (Callon, Lascoumes, and Barthe 2009). The dynamics here owes much to a Deweyan perspective identifying the mechanisms constituting new publics as issues that are not dealt with by existing institutions.[14] Michel Callon's analysis of market in terms of framing/overflowing can be understood in these terms. As Callon explained, the proliferation of markets create "overflows," which results in the formation of new concerned groups, and, potentially, new forms of political organization (Callon 2007).

The joint analytical and political interest for situations characterized by novelty is much more complex to take at face value in the case of nanotechnology—and, indeed, for any domain described in the vocabulary of "emerging technologies." The "novelty" of nanotechnology is perpetually negotiated, used as a resource by its proponents, or questioned. In the United States as well as in Europe, novelty was at the heart of nanotechnology policy. It was contested when the U.S. National Nanotechnology Initiative was created (chapter 1), during discussions about the nature of nano products such as nanosilver (in the United States, chapter 4), or throughout the discussions about the definitions of nanomaterials (in Europe, chapter 5). Constructing "new standards" for a new market, and rethinking the categorization of chemicals so that existing substances become "nano" or not were permanent concerns in the standardization organizations we encountered. Consequently, the language of the "new" entity facing existing modes of problematization cannot account for the constitution of nanotechnology.

For our concern for the democratization possibilities, this means that "novelty" cannot be considered an independent criterion according to which the social scientist could isolate the interesting situations—both at analytical and political levels. This point is made by Michel Callon himself in a 2012 paper, where he argued that "intensive innovation," linking technological and social exploration with renewed democratic constructs, is a particular problematization, and certainly not the only one, or the more valuable (Callon 2012). Callon then suggested that the social scientist ought to multiply the possibilities for alternate problematizations to develop. This leads to a second perspective, which makes pluralism a central objective.

Pluralism

Accounting for various problematizations of nanotechnology and deciphering the variety of articulation between its objects, futures, concerns and publics, one could see pluralism as a guiding principle for both the conduct of analytical work and political engagement. Pluralism is a concept in political theory, and a category of thought originating from pragmatism. William James's "pluralistic universe" points to the philosopher's interest in reality in the making, for variations in the making of things themselves (James 1977). James's pluralism is ontological, in that it seeks to account for the variety of experiences. This resonates well with our use of problematization—a way to account for the construction of objects, futures, concerns, and publics at the same time that they are made collective problems to deal with.

At this stage, one should distinguish between two versions of pluralism, a weak and a strong one. The weak version of pluralism contends that pragmatism invites us to identify various "perspectives" of reality. This stems from a reading of classical pragmatism that focuses on variety across values. For example, discussing the "politics of the pluriverse," a political theorist developed an understanding of William James's political theory based on the pluralism of various orders of worth (Ferguson 2007). Framed this way, pluralism inevitably raises the question of relativism. Hilary Putnam, for instance, refers to Dewey in order to point to the situated objectivity of ethics, "as opposed to an 'absolute' answer to 'perspective-independent' questions" (Putnam 1989, 25). Putnam's pluralist argument leads him to argue for an ethics "without ontology," that is, an ethics that would not refer to a stable and unquestionable Being (Putnam 2004). He is then caught in the problem of relativism, since he wants to retain the objectivity of moral judgment, and the "fundamental values of liberty, autonomy and respects for persons" (Alexander 1993, 376). Putnam solves this problem by considering that objectivity, as in mathematics, is obtained within systems of language. One can thus be "objective without object," and there is no need for a reference to an outside world to sustain ethics' objectivity.[15] Then objectivity is that of the situation within particular language games (in mathematics), or in "practical reasoning" (in ethics) (Putnam 2004, 72). The equivalent of "language games" are thus "frames" or "habits" that define values and acceptable reasoning. A recent book entitled *Pragmatist Ethics* suggests a similar reading of Dewey: the varieties of "frames" and "habits" would determine moral reasoning (Fesmire 2003). A pragmatist ethicist would recognize this variety and locate his own habits.

The political theorists interested in pluralism are concerned with the kind of world we should live in. They attempt to work on the tension between the plurality of values and the need for making a common collective.[16] But in their reflections, they do not discuss what is certainly one of the most interesting outcomes of STS research, namely the interest for ontologies, be they material objects, prospective futures, or political subjects. By contrast, the strong version of pluralism is concerned with "multiple ontologies," as Annemarie Mol would put it (Mol 2002). Pluralism then, is not about focusing the analytical work on variety across values or stable cultural frames, but rather ontological entities. Pluralism is about local answers to situated problems.[17] As such, the analysis of the problematization of nanotechnology that I have been developing could be described as a strong pluralist endeavor. But pluralism itself, even in its strong version, is not enough to account for differences across problematizations. They are not equally stabilized, or equally extended. They are not equally heard, for instance in the international arenas where countries confront each other. At ISO and the OECD, the concern for "science" separating international work separated sovereign policy decisions that make the initiatives of French delegates particularly difficult to hear, as these initiatives attempting to make the social and technical uncertainties of nanotechnology governable are still poorly equipped. The risk of pluralism is to value multiplicity without differentiating among the variety of technical and political formations.

Critical Constitutionalism

Pursuing her inquiry into the mutual production of technical and social order, Sheila Jasanoff has proposed to develop an analysis focusing on "constitutional moments." The expression, borrowed from legal scholar Bruce Ackerman, is defined as follows by Jasanoff:

These are brief periods in which, through the unending contestation over democracy, basic rules of political practice are rewritten, whether explicitly or implicitly, thus fundamentally altering the relations between citizens and the state. To this definition of constitutional change, STS scholars have added an important further dimension: namely, that constitutional moments may encompass the relationship between experts, who underwrite almost all contemporary state action, and citizens, who are collectively subject to the decisions of states. (Jasanoff 2011, 623–624)

From there, one can develop a constitutional analysis—"constitutional" in that it pertains to the allocation of roles and capacity for action within political institutions, and also to the constitutions of governable entities, be they political subjects, future prospects, or poorly identified material objects. Eventually, constitutional analysis proposes to make states, or state-like entities such as the European Union, a topic of empirical study.

Jasanoff's description of "constitutional moments" resonates with what has been the focus of the analysis in this book. The sites of problematization I looked at are windows into constitutional ordering, and those I studied could well be qualified as "constitutional moments." The French national public debate, the experimental form of real-time technology assessment in the United States, the making of "responsible research and innovation" a key component of the European research policy, or the definition of "nanomaterials" in international organizations are situations where the allocation of power within political institutions, the definition of social identities, and the crafting of ontological categories are at stake. They are sites where the nature of the democratic state, or democratic organizations, is questioned, restabilized, or displaced.

The perspective opened by Jasanoff is useful to characterize the type of engagement that this book proposes. The interest for constitutional analysis is both a guiding principle for research work and political engagement. It suggests locating sites of problematization that have constitutional amplitude, and developing their constitutional strength. Indeed, the analytical engagement in the sites of problematization of nanotechnology that I considered made it possible to account for processes that make nanotechnology a constitutional problem, whether related to the transition of the French state, the restabilization of American expertise, the legitimacy of international decision making, or the modalities of European integration.

The approach I adopted is thus a "constitutionalism," in that it is attentive to constitutional problems rather than others. It focuses on sites where both the constitution of social and technical entities and the institutional organization of democratic life are at stake. This constitutionalism is "critical," in a dual sense. It focuses on sites that are situations of trial, and where processes such as extension, purification, or comparison may reproduce or displace constitutional order. These moments are critical in that they display explicit questions raised by the actors involved about the description of the world as it is and as it should be. They may be turning points in the redefinition of accepted problematizations, as they connect the particularities of the issues at stake (such as the engagement of nanotechnology's publics, the definition of its objects, or the government of its anticipated developments) and problems of political philosophy, raising concerns related to the public objectivity of American expertise, the conditions under which the French state can act for the general interest, the sources of democratic legitimacy of European institutions, or the nature of an international negotiation acceptable for all participating countries.

The constitutionalism I propose is also critical in that it displays the normative charge of problematizations, examining how questions about the desirability of constitutional arrangements and the possibility for alternative propositions are voiced and managed to get heard. In doing so, it is necessarily based on the political engagement of the social scientist, who chooses sites and circulates across them, thereby participating in problematization processes. Extension, purification, and comparison are also operations performed by the social scientist, as she inscribes empirical descriptions in longer genealogies, purifies messy empirical fieldwork to display regularities, and uses comparison as an instrument shedding lights on local specificities. This requires that the social scientist adapt her research methodologies so that she can navigate across different constitutional settings, adding reality to certain problems, while also inserting its analysis within the processes that might destabilize dominant problematizations. Critical constitutionalism, then, is critical in that it deploys a wide array of interventions meant to permanently question the world as it is and as it should be.

As he calls for the multiplication of "interferences" between the social scientist and his field of study, STS scholar John Law criticizes the assumption that "there is indeed a common world or collective within which we live and need to live well in together," which grounds a "constitutionalist" approach he argues against. Instead, Law contends that "in practice the world is irredeemably messy" and that "ordering is partial, incomplete, always more or less local, more or less implicit, and therefore more or less disconcerting" (Law 2010, 273, 279). Law sees an irremediable opposition between the multiplication of interferences meant to account for the messiness of the world, and attempts that presuppose the existence of macro order, possibly by introducing procedural criteria to propose desirable democratic paths. By contrast, critical constitutionalism does not presuppose the existence (or the need for) a unique "common world," but offers a direction for the analysis of the constitution of different worlds. It does not use ready-made criteria to evaluate what is democratic and what is not, but neither is it satisfied with the mere multiplication of interferences. It is guided, when conducting research and engaging politically, by the need for accounting for processes of constitutional ordering.

Not all sites of problematization are equally interesting then. It is a matter of research work and political engagement to choose them, and make them relevant for constitutional analysis. Thus, this book has focused on sites where nanotechnology was problematized in such a way that problems for democracy were explicitly raised. By contrast, it did not make laboratories primary sites of investigation. While laboratory studies in the field of nanotechnology are helpful to describe reconfigurations of practices between disciplines, the use of technical instruments and organizational format as coordinating devices, and new concerns for "responsibility" (Hubert 2007; McCarthy and Kelty 2010; Merz and Biniok 2010), the sites I had to focus on connected nanotechnology with constitutional issues, and as such, were all related to the functioning, reproduction, or displacement of political institutions. They were entry points for the study of problematizations of nanotechnology more or less stabilized, more or less open to alternative formations, and interesting in that they were also problematizations of democracy.

Situated Democratization

Critical constitutionalism is a research program that proposes to rethink the study of democracy. It extends the set of sites where the critical study of democracy should be undertaken. It grounds a democratic theory in that it offers a pathway for the empirical description of the sites where democracy is at stake, and for the normative commitments adopted by the researcher. This theory is empirical, in that it does not seek to provide evaluative criteria independently from the particularities of problematic situations. It claims that the core of democratic life deals with the constitution of objects and subjects, and the making and stabilization of public problems. This forces expanding the kinds of sites to reflect upon when considering democratic problems, from political parties and national assemblies to more secluded places such as standardization organizations and science museums. One can then rethink the problem of external democratic criteria introduced earlier in this chapter. Rather than using categories such as "responsibility," "anticipation," or "participation" as external resources for the evaluation of the democratic quality of collective processes, one is thus bound to make them integral components of contemporary democratic life. As such these categories now belong to what is to be analyzed, and, simultaneously, to what can and should be open for contestation.

The approach undertaken here can be read as a proposition for democratization that suggests turning more places into sites of problematization, and draws more connections among them. It calls for an examination of the sites where the possibilities for disagreement occur, and suggests exploring in what ways oppositions are eliminated, or would have the possibility to be heard. As such, the approach developed here does not make any outcome of problematization a democratic construct, but proposes to turn as many sites as possible into places where democracy is at stake. This is the reason why one can conduct that kind of analysis in international organizations and could do so in nondemocratic states. As sociologists Isabelle Thireau and Hua Linshan have shown in their study of Chinese institutions meant to respond to individual or collective protests, the official places where issues are turned into public problems are sites where democratic life might emerge in China, and where, simultaneously, demands are carefully governed so that oppositions remain limited (Thireau and Linshan 2013). As such, these places are crucial sites for potential democratization, and, simultaneously, for the repression of democratic activities.

The analysis of problematization is both modest and ambitious. It is modest in that it is not separated from the social scientist's conduct of research work, from the detailed analysis of problematization processes, and from the circulation of the analyst as she circulates across sites. Yet it is also ambitious in its objective since it proposes, rather than a never-ending examination of local sites, an approach meant to make constitutional constructs apparent. Its critical strength lies in its ability to rethink democratic ordering from its margins, where the basic tenets of democratic life are contested, restabilized, and, in some cases, displaced.

Conclusion

A Passing Fad?

In October 2015, an article in *Time* magazine described nanotechnology as one of the "investment fads and manias" of the past few decades, about which "nobody in the stock market gets excited anymore."[1] This article discussed the dichotomy between the label "nano," hardly used in new companies entering the stock market, and the growing number of "applications and products in a number of industries." This debate echoed what has been central to nanotechnology since the early support programs were launched. Is it pure "hype," a mere marketing scheme for attracting public funds and investor money? Or is the language of the scientific revolution justified? More than fifteen years after the launch of the U.S. National Nanotechnology Initiative, the *Time* article attempted to resolve a similar quandary—and concluded that there existed a range of "applications and products" beyond a label that had been merely designed to attract investors and could not play that role any more.

Rather than accepting the terms of the debate between the "hype" and the "real," the previous chapters have examined the constitution of nanotechnology as a collection of objects, future, concerns, and publics. This assemblage has a history, and takes various forms across the sites encountered in the previous pages. It may well disappear if, for instance, the new "nano" categories in the standardization and regulatory apparatus are not associated any more with a science policy program expecting to bring future innovations to life and endowed with billions of euros and dollars. And as it is constituted and contested, its novelty remains ambivalent.

That nanotechnology articulates the production of material objects with visionary tales about the future, anticipation of collective concerns, and calls for the active participation of various publics is both a sign of its uncertain status and a reason for anyone interested in democratic theory to analyze it in detail. Indeed, it is because each of its components raises democratic issues that problematizing nanotechnology is also problematizing democracy. This is the reason why nanotechnology has provided us with a lens to rethink the terms in which one thinks about contemporary democracy.

A Problem of Political Philosophy

This book has examined two joint (and potentially antagonistic) movements at the heart of contemporary technological developments and, more generally, of many public issues comprising technical components. The first is the making of programs intended to ensure economic progress through technological developments. The second makes the democratization of public choices a means to avoid public controversies. Each of these two movements applies in the case of nanotechnology, and each raises pervasive democratic questions. Who is to determine the direction of technological development, and, indeed, who is to make it a condition for social progress? Under what conditions are technical choices to be accounted for and by whom? What does "democratization" mean in situations where the making of objects, futures, concerns, and publics is away from public scrutiny?

Nanotechnology is a perfect case for reflecting on these questions—asked by the actors involved in its making, by its critics raising objections, and by the sociologist interested in the coproduction of technological programs and democratic order. This book has illustrated how as scholars and public officials make nanotechnology a problem of public participation in science and technology, they struggle to identify concerned publics, and to isolate publics that would be relevant participants from other publics that would be too critical or not engaged enough. The difficulty in identifying the relevant publics for discussions about nanotechnology is directly connected to the difficulty of identifying the relevant topics about which decisions are to be made. As potential issues are noted, the identification of nanotechnology objects is uncertain, and the extent to which collective discussions might direct technology development is far from evident. In the examples discussed in this book, what is at stake when anticipating future issues, governing potential risks, or organizing collective explorations of ethical issues is not well defined. This makes representing, governing, or engaging in nanotechnology complicated tasks, including for those who hope to democratize technology. And these tasks are even more difficult when the time and place of decision making are not settled. Considering the impetus for "public engagement," one wonders: What is being talked about in participatory settings? What are the connections between these discussions and public decision-making? French and U.S. examples described in chapters 2 and 3 showed that these questions do not receive straightforward answers. And participatory settings are but one type of site among many where decisions about nanotechnology are examined. Governing nanotechnology is the objective of regulatory and standardization initiatives, conducted in much more secluded places than any public participation site.

Who should be involved if nanotechnology is to be subjected to democratic treatment? What are the nano-objects that should be governed? How and where should decisions be made? These questions are precisely those that were asked in the empirical sites examined throughout this book. They make nanotechnology a problem of political philosophy, and a complex one—since none of the components of the democratic life can be predetermined. Throughout the previous chapters, this problem has been an empirical lens to consider questions of legitimacy (How are decisions to be made? Under what conditions of accountability?), citizenship (What is the expected role of the public? How are citizens' interventions defined in expertise organizations?), and sovereignty (What is the expected role of the state? How does it act in international arenas such as the European institutions or global organizations?). The approach I have undertaken in this book does not provide criteria to answer these questions in a straightforward manner. Rather, my approach suggests making them topics of analysis for the study of democratic ordering.

Anticipation and Participation

Identifying the objects and subjects of nanotechnology, and the places and procedures of its public treatment raises an essential question for the actors involved as well as the sociologist interested in the relationships between technological development and democracy, and that is: How does democracy function when neither of the components of the democratic life are settled? In the case of nanotechnology the significance of this question is particularly acute—indeed, nanotechnology offers a perfect case for crafting an analytics of democracy attentive to the variety of its components and the diversity of the site where it is at stake. But it is also one case among many others with which contemporary democracies must struggle. Nanotechnology is unique in the articulation it displays between public and private interventions, material constructs and future prospects, calls for democratization and involvement of social science. But its characteristics also echo many contemporary public problems. The association nanotechnology relies on with uncertain objects, futures, concerns, and publics can be identified when considering other emerging technological domains, such as synthetic biology.

More fundamentally, two of the operations nanotechnology is based on are becoming pervasive concerns in contemporary democracies. First, *anticipation* is an operation on which contemporary democracies rely, in order to foresee future threats, be they technical, environmental, or economic. Nanotechnology is construed, in many of the sites we examined, as a problem of anticipating future health or environmental risks, or future social protests. "Responsibility" is heralded as a way of anticipating these potential threats to technological development, and of ensuring the accountability of public institutions (and possibly individual scientists and entrepreneurs). As such, nanotechnology is a telling example of the contemporary evolutions of science and innovation policies. Beyond nanotechnology, anticipation is the rule for the control of industrial activities, the global management of environmental concerns, the planning of economic policies. Future generations become political actors to be taken into account and governments speak in their name when dealing with complex environmental or economic issues.[2] With anticipation comes the risk of the unexpected: as certain options are given greater degrees of reality, others are excluded and might emerge at unexpected times.

Second, *participation* is now a major concern of contemporary democracies. As seen in this book, science policy bodies, expert agencies, and international organizations see public participation as a condition for the successful development of nanotechnology. This is a sign of a more general reproblematization of the modes of public engagement in democratic life, particularly about technological issues. While the implication of local communities in industrial projects, either in collective negotiations or through direct oppositions, is far from new (Graber 2009), the involvement of concerned publics has become an explicit component of industrial, economic, and environmental policies. This makes participation a complicated operation, while the channels of representation are not predetermined (particularly when issues, such as nanotechnology, are elusive), and the actors expected to be involved (from private companies to critical NGOs and interested social scientists) are diverse. With participation goes a concern about exclusion, be that of citizens uninterested in if not unsupportive of science (or walking away from mainstream political parties). In parallel, this makes it necessary to rethink the terms of critique performed by social science, included in various public policies and expected to be "socially relevant."

Anticipation and participation require instruments, such as risk assessment techniques, regulatory initiatives, ethical guidelines, and technologies of democracy. Anticipation and participation are articulated within constitutional organizations that differ across geographical boundaries. This book has showed that nanotechnology associates them with the reproduction of public expertise in the United States, with the current transformations of the French state, with the European integration objective, and with the conduct of international cooperation for the sake of global market making. More generally, one can hypothesize that anticipation and participation are always related to the stabilization of well-known political spaces or redefine others in the face of environmental or economic concerns, or both. And the reason to do so is that they touch on the core of democratic life, namely the possibility for and the government of oppositions. Anticipation is based on the instrumented evaluation of potential threats, to the environment, markets, or technological development, and constructs spaces for the ordering of oppositions that center less on the general framing of issues and more on the technical examination of their components. Participation is about governing protests, but also making new forms of opposition possible, including about the participatory technologies themselves.

Toward an Empirical Democratic Theory

This book has argued that problematizing nanotechnology is also problematizing democracy. Ultimately, the approach I have presented encourages developing a constitutional analysis of democracy by starting the analysis from sites where associations among material objects, future prospects, public concerns, and political subjects are problematized. Nanotechnology required the displacement of the opposition between two perspectives. The first perspective would have insisted on the social and technical uncertainties of the field, understanding it as an experimental territory where reconfigurations are, if not all possible, at least potential and desirable. In contrast, the second perspective would have pointed to the overdetermination of any choices related to technology development. It would have insisted on market forces, and on the combined strength of references to "economic development" and "technological progress." This dichotomy is particularly acute for nanotechnology given the central role of discourses of novelty in the field's development and concurrently, the importance of such massive actors as global companies, powerful states investing in scientific research, and emerging political entities like the European Union. But nanotechnology is certainly not the only field where it plays out. One could think of any other emerging technology programs, or, indeed, many of the current policy interventions, from social policy to economic regulation.

The analysis I propose does not read the problematizations of nanotechnology as either endless opportunities for radical redefinitions of democratic practices, or inevitable reproductions of inescapable external forces. It does not adopt the viewpoint of the technical or social innovator for whom reality could not endlessly be transformed, or that of critics using external causal explanations like "market forces." Instead, it puts forward a critical constitutionalism that locates the processes that stabilize or destabilize constitutional orders, and inscribes these processes in wider temporal evolutions and spatial extensions. And thus, it proposes a genealogical analysis whereby history is a resource to stabilize problematizations—such as the history of nanotechnology in the construction of the U.S. National Nanotechnology Initiative (see chapter 1), and which is attentive to the gradual evolutions of collective order. This analysis makes continuities and discontinuities part of the empirical exploration, and indeed components of what is problematized. In identifying possibilities for change, it forces one to be particularly cautious in the reification of criteria for evaluation of any "democratic quality." Critical constitutionalism, then, proposes to conduct the analysis in the midst of problematization processes, in order to extend the sites where democracy is at stake, while making visible the current transformations of contemporary democracies. In this exploration, nanotechnology has been a lens relevant for bringing into focus the evolutions of contemporary democracies while developing a democratic theory based on the empirical examination of sites where public problems are made, including from the margins of political institutions traditionally associated with the democratic life. One can see here an invitation to make democracy a topic of continuous empirical and critical investigation.

Notes

Prologue

1. Quotes in this paragraph are excerpts from my fieldwork notebook.

2. This process was called *Grenelle de l'Environnement* as a reference to the 1968 Grenelle Agreements, negotiated between the French government and trade unions.

3. Since the mid-1990s, CNDP has been used for planning consultation processes about local infrastructure projects, but is still relatively inexperienced in the field of debates of "general options," which engages entire policies at the national level. A previous case of *débat d'option générale* dealt with nuclear waste policy in the early 2000s.

1 Problematizing Nanotechnology, Problematizing Democracy

1. Unless otherwise specified, quotes in this chapter are excerpts from this interview.

2. The objectives of Nano2Life are presented as follows on the project's website: "Diagnostics—In vivo imaging—In vitro diagnostics, Drug delivery— Nanopharmaceuticals —Nanodevices, Regenerative medicine—Smart biomaterials— Cell therapies, Implants and wearable sensors," www.nano2life.org (accessed January 15, 2011).

3. By considering controversies in the definition of the "nano-ness" of substances, chapter 4 will get back to this point.

4. The argument is made by nanotechnologists themselves (see, e.g., Jain 2004, 2005).

5. This point has been largely developed by the sociology of science. For a recent example of the discussion of the "political qualities" of technological systems, see Barthe 2009.

6. See, for example, Latour 2005 for a presentation of the "sociology of association." The first meaning of the term "political" that Latour proposes is the introduction of new objects, and, thereby, of new associations—in which sense the discovery of a new planet or new materials is inherently political (Latour 2007).

7. Such an evolution has been described as "translational research" (Woolf 2008).

8. Drexler's book *Engines of Creation* became the central reference for nanotechnologists and "futurists" interested in molecular manufacturing (Drexler 1990). The history of the STM has been explored by Davis Baird and Ashley Shew (Baird and Shew 2004). They reconsider its role in the making of the standard history of nanotechnology, and tie it to the strong connections between science and industry in this field. Cyrus Mody wrote a history of the community developing the STM and other laboratory tools that describes how scientific instruments and the community of scientists developing them built "a path to nanotechnology" (Mody 2011).

9. Science journalist Ed Regis narrates the anecdote in an article in the magazine *Wired*. Regis describes how Drexler's attempts to convince Congress to fund a nanotechnology initiative eventually resulted in his own elimination in favor of business interests (Regis 2004; Laurent 2010, 28–32).

10. The discussion of the "grey goo" (that is, an uncontrollable cloud of self-replicating nanomachines) is presented in chapter 11 of Drexler's *Engines of Creation* (see note 8).

11. References to the Lisbon strategy were explicitly introduced in the presentation brochures of the Nano2Life project.

12. For example, in the European Commission's nanotechnology *Action Plan*.

13. Cf., for instance, the case of nanotechnologist Vicki Colvin and her transformation of "responsibility" into a central concern of nanotechnology development (Kelty 2009; McCarthy and Kelty 2010). In France, Aurélie Delemarle described the case of a Commissariat à l'Énergie Atomique (CEA) director, Jean Thermes, who organized nanotechnology research programs in the Grenoble area (Delemarle 2007). At the French national level, the people in charge of the program for the funding of nanotechnology at the National Agency for Research also come from CEA.

14. European Technology Platforms are coordination mechanisms organized at the initiative of the European Commission and scientific actors. They are meant to contribute to the making of the European research policy.

15. www.nano2life.org (accessed January 12, 2011).

16. Interview, Agence Nationale de la Recherche, Paris, April 22, 2009.

17. The case of Nano2Life is an example. In France, one the early policy initiatives in nanotechnology was a network called *Réseau Micro-Nanotechnologies*. In the

United States, the National Nanotechnology Infrastructure Network (NNIN) aims to make infrastructures available for scientific laboratories.

18. This impacted the very organization of the French National Agency for Research (ANR). The Agency replaced two directions of the Ministry for Research in 2005, one devoted to fundamental research and the other to industrial research. The nanotechnology program that ANR manufactured intended to overcome this dichotomy (interview ANR, Paris, April 22, 2009).

19. This perspective is close to that of the "sociology of expectations." But it is less interested in analyzing "retrospecting prospects" and "prospective retrospects" through the study of past and present discourses and the "representations of the future" they convey (Brown and Michael 2003) than in the operationalization of expectations in actual technico-political instruments (see e.g., van Lente and Rip 1998, and Michael 2000 about the performative roles of expectations). The importance of expectations and foresight has been discussed in the case of nanotechnology, although often through a discourse-based analysis (Selin 2007).

20. One can track the integration of futuristic literature back to the making of science policy programs. For example, graphs used in the science fiction-inspired *Age of Spiritual Machines* by Ray Kurzweil, are remobilized in reports of the U.S. National Academy of Science evaluating the National Nanotechnology Initiative (Laurent 2010, 41–42).

21. See, for example, Levidow and Carr 1997. Brian Wynne discussed how the reduction of ethical issues to problems related to risks resulted in public mistrust (Wynne 2001).

22. U.S. Congress, 21st Century Nanotechnology Research and Development Act, P.L. 108–153, S. 189 (H.R. 766), December 2003.

23. See, for instance, a report about ELSA activities in Europe written by a member of the DG Research (Hullmann 2006b). A frequent interpretation is the "wow to yuck" curb, which the public would be supposed to follow in its acceptance of technology. The accuracy of these understandings of public reactions can be questioned (Rip 2006; on the perception of GMOs in Europe, see Marris 2001). This does not change my argument that nanotechnology's publics are integrated in the making of nanotechnology policy.

24. See also Hulmann 2006a, 12, on the need to "take into account" "citizens' expectations and concerns" since "they present an important impact on the acceptance of new technologies on the market and can decide market success or failure."

25. Cf. Kearnes and Wynne 2007 and Macnaghten, Kearnes, and Wynne 2005 for comments about the importance of the deliberation theme in nanotechnology policy, and its consequences for the involvement of social scientists. This latter question is important. I discuss it at further length in chapters 3 and 5.

26. U.S. Congress, 21st Century Nanotechnology Research and Development Act, P.L. 108–153, S. 189 (H.R. 766), December 2003.

27. I will get back to this case in chapter 6.

28. This echoes the analysis of the mobilization of public opinion in the making of a historical narrative (cf. Gaïti 2007 for an example about the creation of the French Fifth Republic). Such an analysis requires a close examination of the performativity of social science for the making of public opinion (Osborne and Rose 1999; Law 2009; for the history of opinion polling, Blondiaux 1998).

29. For a more developed account of this perspective see Kaiser, Kurath, and Maasen 2010.

30. Locating democratic activities as such echoes the perspectives of scholars such as Chantal Mouffe, who contend that the political is to be found in oppositions and antagonisms (Mouffe 2005). The many potential "political" dimensions of nano-technology (cf. the previous section) make the term uneasy to use. My interest lies, at any rate, in the practice of democracy rather than in the making of the "political." Consequently, I will avoid using the term "political," and will focus on the construction of democratic orders.

31. Works looking for causal relationships among unquestioned entities ("social groups," "cultural values," "agenda," ...) are pursued by authors claiming to be "constructionist," or even "post-modernist," who seek to draw connections with the sociology of public problems (Rochefort and Cobb 1994; Bosso 1994).

32. This stream of work was famously introduced by David Snow and has often been reendorsed since then (Snow 1986; Benford and Snow 2000). Works in this approach describe the dynamics of social movement, and the ways (e.g., the defini-tion, or "framing," of the problem) through which a social movement manages to mobilize resources and individuals on a particular topic.

33. Cf. the studies of the "agenda setting function of the mass media" (McCombs and Shaw 1972), which analyze the influence of media on public opinion and/or political agenda (that is, the agenda of institutions known as "political"). Following this perspective, the "agenda setting" stream studies the causal relationships between media activities and the transformation of a question into a public problem.

34. Initiated by Spector and Kitsuse (2001; see Schneider 1985 for an overview), the sociology of public problems studies the way actors define situations as "problem-atic" and contribute to their transformation into public problems. The approach is a self-defined "constructionist" one, which seeks to demonstrate how the nature of the problem and its (material and human) elements as well as the range of possible solutions are constructed.

35. Consider, for instance, the studies of the trajectories of "problems," "solutions" and "political contexts" that Kingdon (1984) proposes. For Kingdon, these three streams are disjointed, have a stable existence, which the analyst can describe, and may cross and/or align in one way or another, thereby transforming an issue into a public one with a range of possible solutions. This vision leaves little room for potential reconfigurations of social identities. It also faces obvious practical difficulties: how to define the "problem" and the "solution"? How to distinguish them from their "political contexts"? These would be tricky questions if one tried to answer them in the case of Nano2Life.

36. As Hilgartner and Bosk (1988) do in order to explain "the rise and falls of social problems."

37. Considered through the lens of the making of collective and individual identities, problematizations could be described as processes that enact "anthropological problems," as used by Collier and Ong to "refer to an interest in the constitution of the social and the biological existence of human beings as an object of knowledge, technical intervention, politics, and ethical discussion" (Collier and Ong 2006, 6).

38. Foucault's argument is precisely that certain issues, such as "the love of boys," are problematized in Ancient Greece as the questions they raise are made explicit in textual materials.

39. Considering problematization as a process to be permanently enacted is a path for the critique of the repressive hypothesis (Foucault 1976). The repressive hypothesis contends that Christianity transformed preexisting sexual behaviors into moral problems to be dealt with by a set of constraining rules. In this perspective, Christianity would have repressed behaviors that were not problematic before it. On the contrary, the study of the problematization of sexual behavior displayed the continuities and small displacements, and the ontological role of rules in the making of sex itself (Foucault 1984, 23). It considers the technologies for regulation of sexuality are produced in the same movement as sexuality itself.

40. I am therefore reluctant to use the term "events," as Rabinow proposes, in stating that "problematizations emerge out of a cauldron of convergent factors (economic, discursive, political, environmental, and the like). Such an emergence is an event. For example, the Greek problematization of pleasure and freedom or the modern problematization of life and governmentality lasted for centuries. Hence, their emergence and articulations is an event of long duration, one that sets events of different scales in motion" (Rabinow 2003, 55). In order not to be caught into the old/new opposition, I prefer not using the idea of event of problematization, which would force me, as it does Rabinow, to distinguish among "scales" of analysis, separating, for instance, problematization from "assemblages" of a shorter temporality (56).

41. Boltanski's critique of the sociologists "fascinated by novelty" grounds his study of the gradual stabilization of the "cadres" category (Boltanski 1987).

42. Another notable example is (Lezaun and Soneryd 2007), in which the authors describe "technologies of elicitation," that is, instruments expected to make publics speak. See also (Felt and Folcher 2010) about "machineries for making publics."

43. This means that the following chapters will not attempt to classify procedures according to their "participatory" or "deliberative" nature. For an example of such a typology, see Rowe and Frewer 2005. See Fiorino 1990 and Rowe and Frewer 2000 for attempts to provide criteria expected to assess the democratic character of citizen participation mechanisms.

44. See, for example, Noortje Marres's critique of the tendency, including in the STS literature, to consider the "place" or "arena" of political confrontation as a given (Marres 2012).

45. See, for example, Latour's infra-reflexivity as a response to the problem of the "reflexivity loops" (Latour 1988a).

46. Whether or not this "traditional historian" exists today is not what matters for my argument here.

47. Deleuze reformulated the argument in his *Foucault*: "Eventually, it is only practice that constitutes the continuity from past to present, or, reciprocally, the way in which present explains the past. Foucault's interviews are entirely part of his oeuvre, since they extend the historical problematization of each of his books to the construction of actual problems, be they about madness, punishment or sexuality" (Deleuze 1986 [2004], 122; my translation).

48. See also Wynne 1996; Woodhouse et al. 2002; Pestre 2004 for discussions of potential "normative" approaches for STS scholars. Chapter 7 will show how the empirical analysis of the problematization of nanotechnology contributes to these discussions.

2   Representing Nanotechnology and Its Publics in the Science Museum

1. Laurent Chicoineau gave me access to the archives of the Grenoble CCSTI. Unless specified otherwise, the quotes in this section are excerpts from this material, which I translated from the French.

2. Interview with Laurent Chicoineau, Paris, May 2009.

3. Interview with Joël Chevrier, a physicist and member of the design team of the nanotechnology exhibit, Grenoble, July 17, 2009.

4. As one of the preparatory documents of the CCSTI stated (translation of "voir par le toucher").

5. Translation of "voir et manipuler l'invisible," CCSTI internal document, July 10, 2006.

6. Interview with Joël Chevrier, Grenoble, July 17, 2009.

7. Daston and Galison's analysis of the move from "representation" to "presentation" in the last, nanotechnology-focused chapter of their book on objectivity directly echoes the type or representation of nanotechnology performed by the nanomanipulator (Daston and Galison 2007).

8. On the importance of pictures in the development of nanotechnology as a science policy program, see Lösch 2006; Ruivenkamp and Rip 2011; Thoreau 2013.

9. See, for example, the current scholarly interest for the "representation of scientific controversies" (Yaneva, Rabesandratana, and Greiner 2009).

10. See, for example, Laurent 2007; chapter 6 will discuss anti-nanotechnology activists' forms of mobilization.

11. Letter of the director of CCSTI to the préfet.

12. Activists speak of the "propaganda" that CCSTI performs. I describe their critical interventions in Grenoble in chapter 6.

13. A national debate about nanotechnology was being held throughout the country at this time (see the prologue and chapters 3 and 6).

14. From a blog entry by Laurent Chicoineau, October 27, 2009.

15. Presentation of the nanotechnology exhibit, *Cité des Sciences* public document.

16. Interview with Laurent Chicoineau, Paris, May 2009.

17. I had access to the project's documents gathered by Laurent Chicoineau at the Grenoble science center.

18. "The Nanodialogue Project: An Integrated Approach to Communication," undated internal document.

19. See, for instance, his work on consensus conference (Joss and Durant 1995) and participatory technology assessment (Joss and Bellucci 2002).

20. Interview with Simon Joss, London, September 2009.

21. Excerpt from the minutes of the Naples workshop.

22. As visible in the email correspondence of the project participants.

23. CCSTI director Laurent Chicoineau voiced the same opinion:

We were always talking about ethics, the whole exhibit was about ethics! Even for visitors, it was too much. There were many people who would come and tell us "but what is this stuff?" They would get out completely threatened. ... They were telling us "we just want to learn things." (...)

There they would tell us they didn't have what they wanted. And what they wanted was information about what nanotechnology was. (interview with Laurent Chicoineau, Grenoble, July 2009; my translation)

24. The expression was used in a preparatory document ("Nano et société: Exposition itinérante et débats publics").

25. In particular, they were concerned about the poor coordination among the various museums where focus groups were conducted. For budgetary reasons, they had to rely on the local teams to organize the focus groups.

26. For instance, the guidelines provided to the museums involved in the project proposed to ask people what they considered to be "risks" and "benefits" of nanotechnology.

27. Nanodialogue final conference, tapescripts, 34.

28. Thus, the other European project on nanotechnology ELSA funded at that time, called *Nanologue*, articulated the organization of "dialogue sessions" involving experts and civil society representatives with the production of scenarios for the future of nanotechnology (Türk et al. 2006).

29. Phone interview, May 2010. Quotes in this paragraph are excerpts from this interview.

30. These topics were previously dealt with by a separate "science in society unit."

31. Interview, EC civil servant, DG Enterprises, Paris, January 2009.

32. Quotes in this paragraph are excerpts from an interview.

33. Phone interview, DG Research and Innovation, May 2010.

34. This expression is used in Bonazzi 2010. It was also developed by Michel Callon and his colleagues to refer to the reinvention of democratic practices in hybrid fora (Callon, Lascoumes, and Barthe 2009). This latter sense is antithetical to the "monitoring of public opinion" (which somehow presupposes that "public opinion" is out there to measure) that the European "technical democracy" proposes. One can hypothesize that Callon's expression made its way into European institutions and was then translated so that it could be articulated with the other components of the European research policy, but a dedicated empirical analysis would be necessary to demonstrate this point.

35. Interview, Washington, DC, March 2009.

36. Nanoscale Science and Engineering Education (NSEE) Program Solicitation NSF 03-044.

37. Interview with Larry Bell, Boston, January 2007.

38. NISE was funded within this program alongside other educational projects, such as "Instructional Materials Development," which "supports development and rigorous testing of prototype instructional materials that promote student learning and interest in nanoscale science, engineering, and technology materials"; and "Nanotechnology Undergraduate Education," which aims to "introduce nanoscale science and technology through a variety of interdisciplinary approaches into undergraduate education."

39. Wendy Crone, *Bringing Nano to the Public. A Collaboration Opportunity for Researchers and Museums*, Washington, DC. A NISE report; later published as Crone 2010.

40. NISE presentation brochure.

41. These quotes are excerpts from an interview with Margaret Glass, Washington, DC, March 27, 2009.

42. In the institutional organization that the NNI constructed, the "implications" part of the program was supposed to be taken care of by different actors, the "Centers for Nanotechnology in Society," to which the examination of nanotechnology ELSA was delegated (see, for example, chapter 5).

43. See, for example, Alpers et al. 2005 for a discussion of the importance of such collaborations elaborated by Boston Museum of Science staff.

44. Quotes in this paragraph are excerpts from the presentation of this exhibit module, available at http://www.nisenet.org/catalog/introduction-nanotechnology-exhibition (accessed July 21, 2016).

45. Crone, *Bringing Nano to the Public*, 6.

46. Boston Museum of Science, "Nanotechnology in Cambridge: What Do You Think?" Background information on nanotechnology, 2008.

47. See, for instance, the case of the citizen panel gathered in 1978 about recombinant DNA, which made Cambridge an exemplary case for citizen involvement in science (Krimsky 1984).

48. NISE network public forum manual, 14.

49. Reich et al. 2011, 86. The forum can even be used as a tool through which even the "societal implications" of nanotechnology can be transmitted to the museum's visitor. This could be done by making "societal implications" a matter of risks and benefits, that is, of other components of a scientific field that could be more or less understood by the public (87).

50. This expression is that of Daniel Yankelovitch (2001), in the title of the book that was mentioned to me by one of the NISE partners involved in the discussions about the forum format.

3   Replicating and Standardizing Technologies of Democracy

1. Above a certain amount of investment, companies are legally required to commission the CNDP to organize a public debate, early enough in the project in order to allow for modifications.

2. A former president of an organizing committee of CNDP debates told me that, during a debate about a liquefied natural gas plant in Dunkerque, in Northern France, he actively tried to rally fishermen who, being illegal immigrants, were reluctant to appear in public.

3. There had been only two "general option" debates before the nanotechnology debate. They had been organized about nuclear waste and transport policy in the South of France—two topics with connections to local considerations.

4. Interview, Paris, October 2009.

5. Commission Nationale du Débat Public, 2010, *Bilan du Débat Public sur la régulation et le développement des nanotechnologies*, Paris, CNDP, 6 (my translation).

6. I heard him using this expression (in French: *le débat réel*) twice in public events where he was asked to present the nanotechnology debate.

7. The government was not legally bound to do so, as the nanotechnology debate was a so-called "débat d'option," that is, related to general policymaking options, and not to a local infrastructure project.

8. This is the subtitle of a 2008 paper about the NCTF (Cobb and Hamlett 2008).

9. Among these publications were Cobb and Hamlett 2008; Cobb 2011; Guston 2011; Hamlett, Cobb, and Guston 2013; and Philbrick and Barandiaran 2009, as well as papers that were more skeptical about the value of the device (e.g., Kleinman and Delborne 2009; Delborne et al. 2011; Powell et al. 2011).

10. The first *conférence de citoyens* was organized in 1998 about GMOs. It was commissioned by the government. Later conferences while keeping the same name, have been organized by various public and private actors, and commented on by social scientists who wrote "instruction books" (*mode d'emploi*) (Bourg and Boy 2000).

11. The Citizens' Technology Forum specialists borrowed the concept of IPE from deliberation theorists.

12. This quote is an excerpt from a paper that describes a previous Citizens' Technology Forum organized by Hamlett and his colleagues.

13. These were expressions used during the online discussions. The transcripts of the online sessions have been made publicly available by the organizers of the NCTF.

14. Indeed, some of the organizers reported that some of the participants did not bother to read on screen when they knew they had a long time to wait before they were allowed to get into the discussions.

15. See Powell et al. 2011 for a critical account of the NCTF focusing on the construction of the "normal" lay citizen.

16. The commissioner of another *conférence de citoyens* told me that many of the panel members she then met were used to participating in conferences and were regularly contacted by the organizing company. A rigorous empirical study would be needed to evaluate this effect, but one can suspect that the growing market for the *conférence de citoyens* goes with the development of a pool of trained participants.

17. These comments reasserted the call for government oversight of potential health risks, increased toxicology research, and development of risk management methodologies, which were considered necessary in order to take into account the potential release of engineered nanoparticles in the environment. The group participated in an initiative launched by the International Center for Technology Assessment (ICTA) that led to the submission to the Environmental Protection Agency (EPA) of a petition that called on the EPA to regulate nanosilver as a pesticide (see chapter 4).

18. Phone interview with Daniel Kleinman, May 2009.

19. Phone interview with Maria Powell, June 2009.

20. The researchers at Madison involved in the two conferences have drawn a comparison of the two events in terms of the identities of the participants and the outcomes of the processes (Kleinman and Delborne 2009).

21. See, for example, Roco 2005. In Europe, the Innovation Union competitiveness report of 2011 discussed the case of biotechnology as an example of failed regulatory and social harmonization, both across the Atlantic and within European member states (European Commission 2011).

22. He was a coauthor of Gavelin, Wilson, and Doubleday 2007.

23. This person was a senior civil servant from the ministry of health and a member of the French delegation to WPN, WPMN, and ISO. She was also involved in the negotiations about the management of nanomaterials within REACH (for explanation of REACH, see chapter 4).

24. The quotes in this paragraph are excerpts from a draft version of the roundtable agenda.

25. The quotes in this paragraph are excerpts from my fieldwork notebook.

26. The quotes in this paragraph are excerpts from the final version of the roundtable agenda.

27. The quotes in this paragraph are excerpts from my fieldwork notebook.

4 Making Regulatory Categories

1. The material I use is drawn from my experience interviewing members of the French delegation at the TC229, participating as a technical expert in the French national organization for standardization's (AFNOR) meetings devoted to nanotechnologies, and attending many public events linked to ISO's activities concerning nanotechnologies.

2. Excerpt from my fieldwork notebook.

3. A fourth working group, dealing with the specification of materials, was added a little later.

4. "Dimensions between approximately 1 and 100 nm are known as the nano-scale" (NNI strategic plan, 2007).

5. "Typically under 100 nm" is from The Royal Society and The Royal Academy of Engineering, *Nanoscience and Nanotechnologies* (2004), 5.

6. "The size range typically between 1 and 100 nm." From Summary Record of the 2nd meeting of the WPMN, April 22, 2007.

7. TC229 Business Plan.

8. Interview, Paris, April 2010. Quotes in this sentence are excerpts from this interview.

9. On the (controversial) history of nanotechnologies support programs in the United States, see, for example, chapter 1, and see McCray 2005 and Laurent 2010, 26–33.

10. "Nanotechnologies—Terminology and Definitions for Nano-objects—Nanoparticle, Nanofibre and Nanoplate" (2008), ISO/TS 27687:2008.

11. Interview, Paris, March 2010.

12. See, for instance, Environmental Protection Agency, *Nanomaterials Stewardship Program: Interim Report* (Office of Pollution Prevention and Toxics, January 2009), a comment on the voluntary approach to nanomaterials declaration pursued by EPA and described in EPA 2004.

13. Environmental Protection Agency, *TSCA Inventory Status of Nanoscale Substances—General Approach* (Washington, DC: EPA, 2006).

14. The oldest patents for anti-bactericide silver were granted in the 1960s (e.g., Werner Degoli, "Silver Ions Bactericidal Compositions," US Patent 3035968, filled August 29, 1960, issued March 1962).

15. EPA, "Pesticide Registration. Clarification for Ion-generating Equipment," *Federal Register* 72, no. 183 (September 21, 2007).

16. NRDC also identified several companies using nanosilver, for applications in medicine or in the food industry (Jennifer Sass, *Nanotechnology's Invisible Threat: Small Science, Big Consequences*, NRDC issue paper, May 2007).

17. Phone interview, Nigel Walker, November 2009.

18. *Petition for Rulemaking Requesting EPA Regulate Nanoscale Silver Products as Pesticides*, *Federal Register* 73, no. 224 (November 19, 2008).

19. Interview with Jaydee Hanson, ICTA, Washington, March 28, 2009.

20. The writers of ICTA's petition were lawyers who knew that the inscription of a new substance in TSCA's inventory might have been easily canceled by a judge if the validity of the criteria was questionable. Consultants specializing in chemical regulation described an example: "An administrative law judge rejected EPA's motion for summary judgment in a TSCA enforcement matter where EPA asserted that submolecular differences between an existing chemical substance and the chemical subject to the enforcement action allowed EPA to treat the latter as "new." *In the Matter of Concert Trading Corp.*, Docket No. TSCA-94-H-19 (July 24, 1997) (Bergeson and Plamondon 2007, 635).

21. They referred to a paper entitled "Interaction of Silver Nanoparticles with HIV-I," published in the *Journal of Nanobiotechnology* (Elechiguerrai al. 2005).

22. SNWG, comments on the SAP background paper (2009), 7.

23. James Delattre, Murray Height, and Rosalind Volpe, "Comments of the Silver Nanotechnology Working Group for Review by the FIFRA Scientific Advisory Panel" (2009), EPA-HQ-OPP-2009-0683; Murray Height, "Evaluation of Hazard and Exposure Associated with Nanosilver and Other Nanometal Oxide Pesticide Products," presentation at SAP-FIFRA, November 3–6, 2009. I also use in this section an email correspondence with Dr. Rosalind Volpe, in charge of the Nanosilver Working Group. Volpe sent me additional documentation about the working group activities.

24. The discussions at the SAP meetings are reported in: EPA, *Meeting Minutes of the FIFRA Scientific Panel Meeting Held November 3–5, 2009 on the Evaluation of Hazards and Exposure of Nanosilver and Other Nanometal Pesticide Products*, Office of Prevention, Pesticides and Toxic Substances Memorandum, 2010. Quotes in this paragraph are excerpts from these meeting minutes.

25. The Safety for Success workshops were held by the European Commission on a yearly basis from 2007 to 2011.

26. Presentation at the Safety for Success workshop, Brussels, October 2–3, 2008.

27. See Sachs 2009 for a comparison between the two approaches.

28. European Chemicals Agency, *Guidance for Identification and Naming of Substances under REACH* (ECHA, 2007), 29.

29. "Nanomaterials in REACH," 6.

30. Conclusion of the first part of the Reach Implementation Project on Nanotechnology (Joint Research Center, REACH Implementation Project. Substance Identification of Nanomaterials, AA N 070307/2009/D1/534733 between DG ENV and JRC, European Commission, March 2011).

31. The European Environmental Bureau, a federation of environmental organizations, has voiced this concern. See also Jouzel and Lascoumes 2011 for a critical perspective on REACH.

32. European Parliament resolution of April 24, 2009, on regulatory aspects of nanomaterials (2008/2208(INI)).

33. Regulation No. 1223/2009 of the European Parliament and of the Council of November 30, 2009 on cosmetic products: article 2, paragraph k.

34. That was the opinion of a member of the French representation in Brussels who was directly involved in the talks that led to this amendment (personal communication, Paris, May 2011). The European Commission later explained that it was impossible to use an "approximate" definition in the European regulation for reasons of legal clarity. From "Types and Uses of Nanomaterials, including Safety Aspects Accompanying the Communication from the Commission to the European Parliament, the Council and the European Economic and Social Committee on the Second Regulatory Review on Nanomaterials," European Commission staff working paper, COM (2012), 572.

35. European Environmental Bureau, *EEB Position Paper on Nanotechnologies and Nanomaterials. Small Scale, Big Promises, Divisive Messages* (February 2008). The 300 nm size limit, in this perspective, is linked to the maximal size for the diffusion across the placenta to the human fetus.

36. Regulation No. 1169/2011 of the European Parliament and of the Council of October 25, 2011, on the provision of food information to consumers.

37. European Commission, Commission Recommendation of October 18, 2011 on the definition of nanomaterial (2011/696/EU). Point 1 urges member states, EU agencies, and economic operators to use this definition. Point 4 defines "particles," "agglomerates," and "aggregates."

38. Scientific Committee on Emerging and Newly Identified Health Risks, *Scientific Basis for the Definition of the Term "Nanomaterials"* (Brussels, 2010).

39. Questions and answers on the European Commission recommendation on the definition of "nanomaterials," http://ec.europa.eu/environment/chemicals/nanotech/faq/questions_answers_en.htm (accessed June 2012).

40. Thus, the European Commission definition states: "In specific cases and where warranted by concerns for the environment, health, safety or competitiveness the number size distribution threshold of 50% may be replaced by a threshold between 1 and 50%."

41. Public consultation concerning the draft decree on the annual declaration of substances at nanoscale, "Observations by the Standardization Committee X457 Nanotechnologies of AFNOR," February 27, 2011. The committee also mentioned the importance of focusing the declaration on substances intentionally produced (this eventually features in the final version of the text).

42. Interview, Chef du bureau des substances chimiques, French ministry of ecology, November 2012.

5 Making Responsible Futures

1. Responsible development refers to the activities undertaken in order to "provide R&D support for knowledge development, identify possible risks for health, environment, and human dignity, and inform the public with a balanced approach about the benefits and potential unexpected consequences" (Roco 2004, 8). See also Roco 2005.

2. William Bainbridge, co-instigator of the NNI, is a fellow of the Institute for Ethics and Emerging Technologies, a "technoprogressist think tank" founded by Nick Bostrom and James Hughes. See Schummer 2005 for a description of the links between transhumanism and nanotechnology programs.

3. This approach to bioethics was famously conceptualized in the 1978 Belmont report: National Commission for the Protection of Human Subjects of Biomedical and Behavioral Research, *The Belmont Report: Ethical Principles and Guidelines for the Protection of Human Subjects of Research* (ERIC Clearinghouse, 1978).

4. This has caused internal dissensions within PCB as well (Jasanoff 2005, 194–196).

5. PCB transcripts: September 7, 2007.

6. Thus, Khushf has been working on research focusing on the construction of living cells (Khushf 2009).

7. In some cases, one can identify the process through which elements of Khushf's positions make their way into policy documents. The following excerpt appears in the 2006 review of the NNI by the National Research Council:

In general, when the social impacts of a new technology are considered, ethics and fundamental research and development are treated as separate. Such an approach keeps facts and values separate, posits risks and benefits that are measurable and scalable, and assumes that uncertainty can be understood and managed scientifically. But because nanotechnology is a potentially disruptive emerging technology, addressing its impacts on society will require a different approach. (NRC 2006, 88)

These words are quoted from Khushf's contribution to the responsible development workshop, and are consistent with his rejection of the fact/value distinction.

8. For example, the excerpt that follows concludes with "responsibility lies with all the stakeholders to make well-informed decisions that will lead to both realizing the benefits and mitigating the risks of nanotechnology" (National Research Council 2006, 92). This could be endorsed by any liberal ethicist.

9. Phone interview, May 2009.

10. Colvin V (2003a), Nanotechnology Research and Development Act of 2003. Testimony before the U.S. House of Representatives Committee on Science, April 9, 2003.

11. Another Center for Nanotechnology in Society was created at UC Santa Barbara.

12. One of the critics of the program became its head and turned the working group into an advisory commission (McCain 2002).

13. Chapter 4 provided other examples, related to the management of potential risks of nano substances and products.

14. See also Fisher 2005 for a critique of the HGP ELSI program by a proponent of RTTA.

15. RTTA has fostered other initiatives. I describe here two of the most characteristic of an otherwise much more diverse program. See, for an introduction, Barben et al. 2008.

16. I have had repeated and fruitful discussions with Erik Fisher over the past few years, and participated in some meetings about his project. I thank him for his support, and for his openness to my external look at his project. For a detailed analysis of the "embedding humanism" project and a critical perspective on its practical consequences, see Thoreau 2013.

17. See Fisher and Mahajan 2006b for a general presentation of the midstream modulation project.

18. See Fisher and Miller 2009 for a presentation of this particular objective.

19. I had the opportunity to participate in research meetings of the individuals involved in Fisher's embeddedness project.

20. François Thoreau examined STIR as an experiment in social sciences, in which a stable protocol is put to test in a number of laboratories in order to demonstrate its value through the circulation of "stories of engagement" produced by "embedded human and social scientists." His work discusses the modalities of the "imperative of reflexivity" that such a project brings to bear on scientists (Thoreau 2013).

21. Ira Bennett, a member of CNS involved in the making of these scenarios, described the process and commented on an example in Bennett 2008.

22. Unless otherwise specified, quotes in this paragraph are excerpts from my field-work notebook.

23. In another project led by Selin, collective discussions were part of the development of scenarios. CNS members proposed an initial description of a nanotechnology-based product developed at the ASU Institute for Biodesign—a device able to measure biomarkers to provide personal analysis of health and potential illness, even before the onset of symptoms. The potential use and development of this device were then discussed during a workshop by bioethicists, sociologists, political scientists, journalists, and physicists. Eventually four scenarios were produced, which proposed four different versions of the technology and its use through narratives involving the device (e.g., a young man uncertain about whether and how to use an illness tracking device that had supposedly become available at little cost).

24. The person in charge of nanotechnology at the Directorate-General for Research and Innovation advanced the "fragmentation" argument (Hullmannn 2006b).

25. "Decision No 1513/2002/EC of the European Parliament and of the Council of 27 June 2002 concerning the Sixth Framework Programme of the European Community for research, technological development and demonstration activities, contributing to the creation of the European Research Area and to innovation (2002–2006)"; "Decision No 1982/2006/EC of the European Parliament and of the Council of 18 December 2006 concerning the Seventh Framework Programme (2007–2013)."

26. One could discuss the analytical differences of "principles" and "values." As they are used almost interchangeably in the European institutions, I do not address this terminology issue.

27. Interview, Washington, DC, October 2009.

28. European Commission, FP7 Cooperation Work Program 2010. Theme 4: Nano-sciences, Nanotechnologies, Materials and New Production Technologies, European Commission C(2009) 5893, 29 July 2009, 8.

29. The EGE is asked by the European legislation to "establish close links with the Commission departments involved in the topic the Group is working on" (Commission Decision, 11 May 2005, art. 4[2]). The regulation defining the Seventh Framework Programme stated that the "opinions of the European Group on Ethics in

Science and New Technologies are and will be taken into account" (Regulation (EC) No 1906/2006 of the European Parliament and of the Council of 18 December 2006 laying down the rules for the participation of research centers and universities in actions under the Seventh Framework Programme and for the dissemination of research results (2007–2013), art. 30).

30. The *Opinion* proposed to create a European network on ethics (European Group on Ethics 2007, 62). When considering human enhancement, the EGE referred to a previous *Opinion* it had released on ICT implants, in which it proposed that the "enhancement of physical and mental capabilities" (a topic central to discussions about nanotechnology) should be limited to a few well-defined cases, such as the "improvement of health prospects," for instance to "enhance the immune system to be resistant to HIV," and proposed to ban other applications, including "changing the identity, memory, self-perception and perception of others," or "coercion towards others who do not use such a device" (European Group on Ethics 2005, 33–34).

31. Phone interview with DK, in charge of Ethics Review at DG Research, April 22, 2009.

32. The expression "committee shopping" is used to characterize the way in which research projects choose to follow the guidelines of the least constraining national ethics committee among those of the member states (Tschudin 2001).

33. Presentation on "Responsible Innovation" by the Science in Society Unit, Vienna, June 2012.

6   Mobilizing against, Mobilizing within Nanotechnology

1. The expression is used in the corporate communication of Minatec, as in the first "Lettre Minatec" of 2001.

2. Agence d'Etudes et de Promotion de l'Isère, *Les biotechnologies: Une convergence de disciplines pour les sciences de la vie* (2003).

3. This is a quote of Geneviève Fioraso, a city councilor and later minister for research in the French government, talking about Nanobio in a session of the municipal council (Conseil municipal de Grenoble, "Pôle d'innovation nanobio-technologies 'Nanobio,' Convention de fonctionnement du pôle," Grenoble, November 27, 2006).

4. André Vallini, "Discours pour l'inauguration de Minatec," June 2, 2006.

5. The expression was used by a CEA official, interview, Grenoble, January 15, 2007. Jean Therme's strategy has been described as that of an "institutional entrepreneur" (Mangematin et al. 2005; Delemarle 2007).

6. Interview, CEA official, Grenoble, January 15, 2007.

7. The example of Grenoble was central in a report to the Prime Minister about competitiveness written by MP Christian Blanc (2004).

8. This is a claim the activists made. I did not find other examples of a public demonstration against nanotechnology before this date.

9. "Minatec: inauguration policière," *Opposition Grenobloise aux Nécrotechnologies*, http://www.piecesetmaindoeuvre.com/spip.php?article77 (accessed December 2, 2010).

10. The material for the analysis of PMO is based on observations of activists' meetings, interviews in Grenoble, and texts written by activists (available on the group's website www.piecesetmaindoeuvre.com, or distributed during meetings).

11. Excerpt from the radio program *Là bas si j'y suis*, France Inter, "Nanotechnologies: refus de modernité ou d'inhumanité?," June 2, 2006.

12. "Louis Néel à Grenoble: La liaison militaro-scientifique" (Louis Néel at Grenoble: The military-scientific link) is the title of one of the pieces written by Simple Citoyen and published online in 2004 (http://www.piecesetmaindoeuvre.com/IMG/pdf/L._Neel.pdf [accessed July 19, 2016]).

13. Alain Carignon, mayor of Grenoble from 1983 to 1995, was involved in a corruption scandal and sentenced to jail for five years.

14. Interview with J. Caunes, councilor at *La Métro*. Caunes was a councilor from a minority group who advocated "public dialogue."

15. PMO, "La Métro tente de recruter Pièces et Main d'Oeuvre," http://www.piecesetmaindoeuvre.com/spip.php?article56 (accessed December 12, 2010).

16. Vivagora's internal document.

17. She used PMO's anonymous email address.

18. PMO, "La Métro tente de recruter Pièces et Main d'Oeuvre."

19. The phrase was used by a Grenoble scientist in biomedicine (interview, Grenoble, January 2007).

20. The expression was used in several interviews with local scientists and officials.

21. Interview with Hélène Mialet, city councilor, January 2007.

22. Similar problems occurred for the social scientists asked to advise the local administrative bodies in Grenoble. The experience was reflected on by two of the authors of the report they wrote (Joly et al. 2005), who discussed the practical difficulties of the abstract "upstream engagement" objective (Joly and Kaufmann 2008).

23. Interview with a city councilor, Grenoble, January 2007.

24. I saw the president of the CNDP doing this presentation in two conferences in 2010. Other presentations were reported to me by the actors I interviewed.

25. The quantitative argument is not that certain. According to one of the members of the expert group assisting CNDP in the organization of the debate, CNRS (a public research body) told heads of laboratories to ask their researchers to come and fill up the rooms.

26. These expressions are excerpts from an interview with a city councillor, Grenoble, January 2007.

27. Excerpt from the radio program *Là bas si j'y suis*, France Inter, "Nanotechnologies: refus de modernité ou d'inhumanité?," June 2, 2006.

28. Interview, university student, Grenoble, January 2007.

29. Indymedia Grenoble is one of them. It is part of a global network (Independent Media Center, Indymedia) created after the demonstrations in Seattle in 1999 and devoted to independent information on an anti-globalization agenda (Morris 2004).

30. For instance, the Canadian NGO ETC Group has been advocating a moratorium on "nanomaterials" since the early 2000s.

31. Quotes in this section are excerpts from the notes I took while sitting in this meeting. This meeting had been announced on a website conceived as a platform for the opposition against the CNDP debate.

32. *Rapport final du débat public*, 5 (my translation).

33. I sat in this meeting, which had also been announced online.

34. PMO, "Memento Malville," 2005.

35. PMO, "La Métro tente de recruter Pièces et Main d'Oeuvre.

36. Ibid. "Technical democracy" is an expression used in Callon et al. 2009. PMO members are familiar with this academic reference.

37. I use a variety of empirical material in this section: notes taken during public meetings and Vivagora's internal meetings, discussions with Vivagora's members, and public and internal documents related to Vivagora's activities. Unless otherwise specified, quotes in this section are excerpts from my fieldwork notebooks.

38. The Nanoforum format was later replicated on synthetic biology, after a report I coauthored with Pierre-Benoît Joly, Claire Marris, and Douglas Robinson had suggested to do so (Joly et al. 2011). The first meeting of the Nanoforum-like device was interrupted by activists (see an account of the event by Morgan Meyer, www.csi .mines-paristech.fr/blog/?p=124 [accessed August 21, 2016]). That this replication

was not as consensual as the original Nanoforum is a sign that the French state is not equipped to deal with the uncertain issues it seeks to include in its domains of expertise (see a discussion of this point in chapter 7).

39. "Projets 2008–2009," internal document, Vivagora.

40. "Nanotechnologies, osez mettre en débat les finalités," *Le Monde*, February 18, 2010.

41. Interview, Washington, DC, March 29, 2010.

42. Thus, participants in a 2010 EEB meeting on nanotechnology, which I attended, insisted on the difficulties for NGOs to attend ISO meetings because of financial costs, and be numerous enough to participate in the works of multiple technical committees.

7   Democratizing Nanotechnology?

1. Thus, Marcus insists on the possibility offered by a multi-site ethnography for the multiplicity of the forms of intervention of the social scientist, who can then engage in "activism," but "activism quite specific and circumstantial to the conditions of doing multi-sited research itself. It is a playing out in practice of the feminist slogan of the political as personal, but in this case it is the political as synonymous with the professional persona and, within the latter, what used to be discussed in a clinical way as the methodological" (Marcus 1995, 113). Marcus goes on by providing the example of an anthropologist studying AIDS, who is, in turn, "an AIDS volunteer at one site, a medical student at another, and a corporate trainee at a third." Michiel van Oudheusden and I saw the political value of an "experimental normativity" in the multiplicity of forms of intervention, and the specificity of the academic role in the possibility of shifting from site to site, and experimenting with a variety of positions (van Oudheusden and Laurent 2013). In this chapter, I want to connect this interest in the multiplicity of modes of intervention with the need for reconstruction in order to display and critique the problematizations of nanotechnology.

2. This is, however, Law's approach, as he develops his critique of "constitutionalism," to which I will return in the final section of this chapter (Law 2010).

3. This is a direct follow-up to the scholarly works that have demonstrated that no experiment can be conducted without simultaneously shaping legitimate audiences, and carefully crafting public spaces (Shapin and Schaffer 1985). This suggests discussing the notion of "political experiments" that has become trendy among STS scholars interested in political theory. One can demonstrate that the study of experiments (whether "scientific" or "political") requires that one analyze the spaces within which they are valued. When the experimental sites examined are those where democracy is problematized, they often become empirical entry points for

the study of institutionalized political spaces, such as the French state in transition (Laurent 2016b).

4. The famous report written by Vannevar Bush, *Science: The Endless Frontier*, advocated for federal support for basic research (Bush [1945] 1960)

5. The Woodrow Wilson Center and the ICTA were among them. The Environmental Defense Fund was also particularly active. See Hess 2010 for a detailed account on the evolution of the funding of EHS research for nano substances and products within the NNI.

6. U.S. Government Accountability Office, *Nanotechnology: Better Guidance Is Needed to Ensure Accurate Reporting of Federal Research Focused on Environmental, Health, and Safety Risks* (Washington, DC: GAO, 2008).

7. Existing scientometrics in nanotechnology differentiates "toxicology" fields from other domains of scientific activities (Youtie et al. 2011).

8. Marc Mortureux, "L'indépendance de l'expertise ne repose pas sur des "moines-chercheurs," *Le Monde*, December 19, 2012.

9. See, among other examples, Porter 1996.

10. *Innovation Union Competitiveness Report* (Brussels: European Commission, 2011), 12.

11. See Trubeck and Trubeck 2005 for a discussion of the experimentalist perspective on the construction of European social policy. See Szyszczak 2006 for a description of the open method of coordination as an instrument in "experimental governance," in a perspective defined by Sabel and Zeitlin (2003). The 2005 nanotechnology *Action Plan* of the European Commission explicitly refers to the open method of coordination as a way of making member states follow the Lisbon strategy: "in line with the subsidiarity principle, the Commission considers the 'Open Method of Coordination' to be an appropriate way to proceed with the use of information exchange, indicators, and guidelines" (European Commission 2005).

12. I participated in this roundtable as I was working at the OECD WPN. I use here the field notes I took during this event.

13. See Kurath 2009. Kurath's paper has provoked some discussions about the validity of such criteria, and the empirical and normative strength of mode 2 in the first place (Rip 2010).

14. For Dewey, problematization is a matter of novelty, as it occurs whenever a situation has "needs," that is, faces issues not dealt with by existing arrangements (Rabinow 2003). STS work interested in issue politics tends to reproduce this binary opposition between "new" issues opposed to "old" institutions (Marres 2007). See also the invitation for "innovations in politics" that would produce "sur-

prises" in otherwise well-known existing political constructions that Emilie Gomart and Maarten Hajer proposed (Gomart and Hajer 2003).

15. Thus, Putnam said: "If I am asked to explain how ethical knowledge is possible at all in "absolute" terms, I have no answer" (Putnam 1989, 22).

16. See, for example, Dryzek and Niemeyer 2006.

17. The equivalence between pluralism and pragmatism has been criticized by political theorists such as Robert Talisse (Talisse and Aikin 2005). My description of "strong pluralism" would fit with what is, for Talisse, pragmatism.

Conclusion

1. "Here's Why Nobody Is Talking about Nanotech Anymore," *Time*, October 9, 2015.

2. See, e.g., Lemoine and Lelann 2012 about the call to future generations in the government of public spending in Europe. One can identify the same reference to future generations in all domains related to sustainability, whether they are related to environmental or economic concern.

References

Ach, Johann, and Ludwig Siep. 2006. *Nano-Bio-Ethics: Ethical Dimensions of Nanobiotechnology*. Münster: LIT Verlag.

Alexander, Thomas M. 1993. "John Dewey and Moral Imagination: Beyond Putnam and Rorty toward a Postmodern Ethics." *Transactions of the Charles S. Pierce Society* 29 (3): 369–400.

Alpers, Carol Lynn, Jacqueline Isaacs, Carol Barry, and Glen Miller. 2005. "Nano's Big Bang: Transforming Engineering Education and Outreach." Paper presented at the American Society for Engineering Education Annual Conference & Exposition, Portland, OR, June 12–15.

Auffan, Mélanie, Jérôme Rose, Jean-Yves Bottero, Gregory Lowry, Jean-Pierre Jolivet, and Mark Wiesner. 2009. "Towards a Definition of Inorganic Nanoparticles from an Environmental, Health and Safety Perspective." *Nature Nanotechnology* 4:634–641.

Baird, Davis, and Ashley Shew. 2004. "Probing the History of Scanning Tunneling Microscopy." In *Discovering the Nanoscale*, ed. David Baird, Alfred Nordmann, and Joachim Schummer, 145–156. Amsterdam: IOS Press.

Barben, Daniel, Erik Fisher, Cynthia Selin, and David Guston. 2008. "Anticipatory Governance of Nanotechnology: Foresight, Engagement and Integration." In *The Handbook of Science and Technology Studies*, ed. Edward J. Hackett, Olga Amsterdamska, Michael Lynch, and Judith Wajcman, 979–1000. Cambridge, MA: MIT Press.

Barry, Andrew. 1993. "The European Community and European Government: Harmonization, Mobility and Space." *Economy and Society* 22 (3): 314–326.

Barry, Andrew. 1999. "Demonstrations: Sites and Sights of Direct Action." *Economy and Society* 28 (1): 75–94.

Barry, Andrew. 2001. *Political Machines: Governing a Technological Society*. London: Athlone Press.

Barthe, Yannick. 2009. "Les qualités politiques des technologies: Irréversibilité et réversibilité dans la gestion des déchets nucléaires." *Tracés* 16 (1): 119–137.

Bell, Larry. 2008. "Engaging the Public in Technology Policy: A New Role for Science Museums." *Science Communication* 29 (3): 386–398.

Benamouzig, Daniel, and Julien Besançon. 2005. "Administrer un monde incertain: Les nouvelles bureaucraties techniques: Le cas des agences sanitaires en France." *Sociologie du Travail* 47 (3): 301–322.

Benford, Robert, and David Snow. 2000. "Framing Processes and Social Movements: An Overview and Assessment." *Annual Review of Sociology* 26:611–639.

Bennett, Ira. 2008. "Developing Plausible Nano-enabled Products." In *Yearbook for Nanotechnology in Society*, ed. Erik Fisher, Cynthia Selin, and Jameson Wetmore, 149–155. Dordrecht: Springer.

Bennett, Ira, and Daniel Sarewitz. 2006. "Too Little, Too Late? Research Policies on the Societal Implications of Nanotechnology in the United States." *Science as Culture* 15 (4): 309–325.

Bennett, Tony. 1995. *The Birth of the Museum: History, Theory, Politics*. London: Routledge.

Bensaude-Vincent, Bernadette. 2000. *L'opinion publique et la science: À chacun son ignorance*. Paris: La Découverte.

Bensaude-Vincent, Bernadette. 2001. "The Construction of a Discipline: Materials Science in the United States." *Historical Studies in the Physical and Biological Sciences* 31 (2): 222–248.

Bensaude-Vincent, Bernadette. 2004. "Two Cultures of Nanotechnology." *HYLE—International Journal for Philosophy of Chemistry* 10 (2): 65–82.

Berger, François, Sjef Gevers, Ludwig Siep, and Klaus-Michael Weltring. 2008. "Ethical, Legal and Social Aspects of Brain-implants Using Nano-scale Materials and Techniques." *NanoEthics* 2:241–249.

Bergeson, Lynn, and Joseph Plamandon. 2007. "TSCA and Engineered Nanoscale Substances." *Nanotechnology Law and Business* 4 (1): 617–640.

Berube, David. 2006. *Nano-Hype: The Truth behind the Nanotechnology Buzz*. New York: Prometheus Books.

Bimber, Bruce. 1996. *The Politics of Expertise: The Rise and Fall of Technology Assessment*. Albany: SUNY Press.

Blanc, Christian. 2004. *Pour un écosystème de la croissance, Rapport au Premier Ministre*. Paris: Assemblée Nationale.

Blondiaux, Loïc. 1998. *La Fabrique de l'Opinion: Une histoire sociale des sondages*. Paris: Seuil.

Boltanski, Luc. 1979. "Taxonomie sociale et luttes de classes: La mobilisation de la 'classe moyenne' et l'invention des cadres." *Actes de la Recherche en Sciences Sociales* 29:75–106.

Boltanski, Luc. 1987. *The Making of a Class: Cadres in French Society.* Cambridge: Cambridge University Press.

Bonazzi, Matteo, ed. 2007. "Working Paper Resulting from the Workshop on Strategy for Communication Outreach in Nanotechnology (Brussels, 6th February 2007)." Brussels: European Commission.

Bonazzi, Matteo. 2010. *Communicating Nanotechnology: Why? To Whom? Saying What? and How?* Luxembourg: Publications Office of the European Union.

Borraz, Olivier. 2007. "Governing Standards: The Rise of Standardization Processes in France and in the E.U." *Governance: An International Journal of Policy, Administration and Institutions* 20 (1): 57–84.

Bosso, Christopher. 1994. "The Contextual Basis of Problem Definition." In *The Politics of Problem Definition*, ed. David Rochefort and Roger Cobb, 182–203. Lawrence: University Press of Kansas.

Bostrom, Nick. 2003. *The Transhumanism FAQ: A General Introduction.* The World Transhumanist Association. http://www.nickbostrom.com/views/transhumanist.pdf (accessed March 24, 2016).

Boullier, Henri, and Brice Laurent. 2015. "La precaution réglementaire. Un mode européen de gouvernement des objets techniques." *Politique Européenne* 49:30–53.

Bourdieu, Pierre. 1980. *Le sens pratique.* Paris: Minuit.

Bourg, Dominique, and Daniel Boy. 2000. *Conférences de Citoyens: Mode d'emploi.* Paris: Descartes.

Bowker, Geoffrey, and Leigh Star, eds. 1999. *Sorting Things Out: Classification and Its Consequences.* Cambridge, MA: MIT Press.

Brown, Nik, and Mike Michael. 2003. "A Sociology of Expectations: Retrospective Prospects and Prospective Retrospects." *Technology Analysis and Strategic Management* 15 (1): 3–18.

Bush, Vannevar. [1945] 1960. *Science: The Endless Frontier.* Washington, DC: National Science Foundation.

Callon, Michel. 1980. "Struggles and Negotiations to Define What Is Problematic and What Is Not: The Socio-logic of Translation." In *The Social Process of Scientific Investigation: Sociology of the Sciences Yearbook*, ed. Karin Knorr Cetina and Aaron Cicourel, 197–220. Dordrecht: D. Reidel.

Callon, Michel. 1986. "Some Elements of a Sociology of Translation. Domestication of the Scallops and the Fishermen of St Brieuc Bay." In *Power, Action and Belief: A New Sociology of Knowledge*, ed. John Law, 189–203. London: Routledge and Kegan Paul.

Callon, Michel. 1998. "Des différentes formes de démocratie technique." *Annales des mines* 9: 63–73.

Callon, Michel, ed. 1998. *The Laws of the Markets*. London: Blackwell.

Callon, Michel. 1999. "Ni sociologue engagé, ni sociologue dégagé: La double stratégie de l'attachement et du détachement." *Sociologie du Travail* 41 (1): 65–78.

Callon, Michel. 2004. "Europe Wrestling with Technology." *Economy and Society* 1 (33): 121–134.

Callon, Michel. 2007. "An Essay on the Growing Contribution of Economic Markets to the Proliferation of the Social." *Theory, Culture & Society* 24 (7/8): 139–163.

Callon, Michel. 2008. "Economic Markets and the Rise of Interactive Agencements: From Prosthetic Agencies to Habilitated Agencies." In *Living in a Material World: Economic Sociology Meets Science and Technology Studies*, ed. Trevor Pinch and Richard Swedberg, 29–56. Cambridge, MA: MIT Press.

Callon, Michel. 2009. "Civilizing Markets: Carbon Trading between *In Vitro* and *In Vivo* Experiments." *Accounting, Organizations and Society* 34 (3–4): 535–548.

Callon, Michel. 2012. "Quel rôle pour les sciences sociales face à l'emprise grandissante du régime de l'innovation intensive?" *Cahiers de Recherche Sociologique* 53:121–165.

Callon, Michel, Pierre Lascoumes, and Yannick Barthe. 2009. *Acting in an Uncertain World: An Essay on Technical Democracy*. Cambridge, MA: MIT Press.

Callon, Michel, and Bruno Latour. 1981. "Unscrewing the Big Leviathan: How Actors Macrostructure Reality and How Sociologists Help Them to Do So." In *Advances in Social Theory and Methodology: Toward an Integration of Micro and Macro-Sociologies*, ed. Karin Knorr Cetina and Aaron Cicourel, 277–303. London: Routledge and Kegan Paul.

Caron, François. 2000. "Le dialogue entre la science et l'industrie à Grenoble." *Revue pour l'histoire du CNRS* 2:44–52.

Chang, Robert, and Rob Semper, eds. 2004. *Opportunities and Challenges of Creating an Infrastructure for Public Engagement in Nanoscale Science and Engineering*. Report of a National Science Foundation Workshop, September 2–3, Washington, DC.

Chilvers, Jason. 2010. *Sustainable Participation? Mapping Out and Reflecting on the Field of Public Dialogue on Science and Technology*. Harwell, England: Sciencewise Expert Resource Centre.

Cobb, Michael. 2011. "Creating Informed Public Opinion: Citizen Deliberation about Nanotechnologies for Human Enhancements." *Journal of Nanoparticle Research* 13 (4): 1533–1548.

Cobb, Michael, and Patrick Hamlett. 2008. "The First National Citizens' Technology Forum on Converging Technologies and Human Enhancement: Adapting the Danish Consensus Conference in the USA." Paper presented at the 10th Conference on Public Communication of Science and Technology, Malmö, Sweden, June 25–27.

Collier, Stephen, and Aihwa Ong. 2006. "Global Assemblages, Anthropological Problems." In *Global Assemblages: Technology, Politics, and Ethics as Anthropological Problems*, ed. Aihwa Ong and Stephen Collier, 3–21. London: Blackwell.

Crone, Wendy. 2010. "Bringing Nano to the Public: A Collaboration Opportunity for Researchers and Museums." *Journal of Nano Education* 2 (1–2): 102–116.

Dab, William, and Danielle Salomon. 2013. *Agir face aux risques sanitaires*. Paris: P.U.F.

Daston, Lorraine, and Peter Galison. 2007. *Objectivity*. New York: Zone Books.

Davies, Sarah, Phil Macnaghten, and Matthew Kearnes, eds. 2009. *Reconfiguring Responsibility: Lessons for Public Policy, Part 1 of the DEEPEN Report*. Durham: Durham University.

Delborne, Jason, Ashley Anderson, Daniel Kleinman, Mathilde Colin, and Maria Powell. 2011. "Virtual Deliberation? Prospects and Challenges for Integrating the Internet in Consensus Conferences." *Public Understanding of Science* 20 (3): 367–384.

Delemarle, Aurélie. 2007. "Les leviers de l'action de l'entrepreneur institutionnel: Le cas des micro et nanotechnologies et du pôle de Grenoble." Unpublished PhD thesis. Paris: Ecole des Ponts.

Deleuze, Gilles. [1986] 2004. *Foucault*. Paris: Minuit.

Dewey, John. [1927] 1991. *The Public and Its Problems*. Athens, OH: Ohio University Press/Swallow Press.

Dratwa, Jim. 2012. "Representing Europe with the Precautionary Principle." In *Reframing Rights: Bioconstitutionalism in the Genetic Age*, ed. Sheila Jasanoff, 263–286. Cambridge, MA: MIT Press.

Drexler, Erik. 1990. *Engines of Creation*. London: Fourth Estate.

Drexler, Erik. 2004. "Nanotechnology: From Feynman to Funding." *Bulletin of Science, Technology & Society* 24 (1): 21–27.

Dryzek, John, and Simon Niemeyer. 2006. "Reconciling Pluralism and Consensus as Political Ideals." *American Journal of Political Science* 50 (3): 634–649.

Duncan, Carol. 1995. *Civilizing Rituals: Inside Public Art Museums*. London: Routledge.

Durant, John. 2004. "The Challenges and Opportunities of Presenting 'Unfinished Science." In *Creating Connections: Museums and the Public Understanding of Current Research*, ed. David Chittenden, Graham Farmelo, and Bruce Lewenstein, 47–60. Walnut Creek: Altamira Press.

Elechiguerrai, Jose Luis, Justin Burt, Jose Morones, Alejandra Camacho-Bragado, Xiaoxia Gao, Humberto Lara, and Jose Yacaman. 2005. "Interaction of Silver Nanoparticles with HIV-I." *Journal of Nanobiotechnology* 3 (6): 1–10.

EPA (Environmental Protection Agency). 2004. *Concept Paper for the Nanomaterial Stewardship Program under TSCA*. EPA-HQ-OPTT-2004-0122-058.

European Commission. 2004. *Communication: Toward a European Strategy for Nanotechnology*. Luxembourg: Office for Official Publications of the European Communities.

European Commission. 2005. *Nanosciences and Nanotechnologies: An Action Plan for Europe 2005–2009*. Brussels: European Commission.

European Commission. 2007. *The European Research Area: New Perspectives. Inventing Our Future Together*. Luxembourg: Office for Official Publications of the European Communities.

European Commission. 2009. *Commission Recommendation on a Code of Conduct for Responsible Nanosciences and Nanotechnologies Research & Council Conclusions on Responsible Nanosciences and Nanotechnologies Research*. Luxembourg: Office for Official Publications of the European Communities.

European Commission. 2011. *Innovation Union Competitiveness Report: 2011 Edition*. Luxembourg: Publications Office of the European Union.

European Commission. 2012. *Responsible Research and Innovation: Europe's Ability to Respond to Societal Challenges*. Luxembourg: Publications Office of the European Union.

European Commission. 2013. *Options for Strengthening Responsible Research and Innovation*. Report of the Expert Group on the State of Art in Europe on Responsible Research and Innovation. Luxembourg: Publications Office of the European Union.

European Group on Ethics in Science and New Technologies to the European Commission. 2000. *General Report on the Activities of the European Group on Ethics in Science and New Technologies to the European Commission, 1998–2000*. Luxembourg: Office for Official Publications of the European Communities.

European Group on Ethics in Science and New Technologies to the European Commission. 2005. *Opinion on the Ethical Aspects of ICT Implants in the Human Body.* Luxembourg: Office for Official Publications of the European Communities.

European Group on Ethics in Science and New Technologies to the European Commission. 2007. *Opinion on the Ethical Aspects of Nanomedicine.* Luxembourg: Office for Official Publications of the European Communities.

European Technology Platform on Nanomedicine. 2005. *Vision Paper and Basis for a Strategic Research Agenda for Nanomedicine.* Luxembourg: Office for Official Publications of the European Communities.

Evans, John. 2000. "A Sociological Account of the Growth of Principlism." *Hastings Center Report* 30 (5): 31–38.

Evans, John H. 2002. *Playing God? Human Genetic Engineering and the Rationalization of Public Bioethical Debate.* Chicago: University of Chicago Press.

Felt, Ulrike, and Maximilian Fochler. 2010. "Machineries for Making Publics: Inscribing and De-scribing Publics in Public Engagement." *Minerva* 48 (3): 219–238.

Felt, Ulrike, and Brian Wynne. 2007. *Taking European Knowledge Society Seriously. Report of the Expert Group on Science and Governance to the Science, Economy and Society Directorate, DG Research, European Commission.* Luxemburg: Office for Official Publications of the European Communities.

Ferguson, Kennan. 2007. *William James: Politics in the Pluriverse,* ed. Morton Schoolman Lanham, MD: Rowman and Littlefield.

Fesmire, Steven. 2003. *Pragmatist Ethics: John Dewey and Moral Imagination.* Bloomington: Indiana University Press.

Fiorino, Daniel. 1990. "Citizen Participation and Environmental Risk: A Survey of Institutional Mechanisms." *Science, Technology & Human Values* 15 (2): 226–243.

Fisher, Erik. 2005. "Lessons Learned from the Ethical, Legal and Social Implications Program (ELSI): Planning Societal Implications Research for the National Nanotechnology Program." *Technology in Society* 27:321–328.

Fisher, Erik. 2007. "Ethnographic Invention: Probing the Capacity of Laboratory Decisions." *NanoEthics* 1 (2): 155–165.

Fisher, Erik, and Roop Mahajan. 2006a. "Contradictory Intent? US Federal Legislation on Integrating Societal Concerns into Nanotechnology Research and Development." *Science & Public Policy* 33 (1): 5–16.

Fisher, Erik, and Roop Mahajan. 2006b. "Midstream Modulation of Nanotechnology Research in an Academic Laboratory." Paper presented at the IMECE2006 ASME International Mechanical Engineering Congress and Exposition, Chicago, November 5–10.

Fisher, Erik, Roop Mahajan, and Carl Mitcham. 2006. "Midstream Modulation of Technology: Governance from Within." *Bulletin of Science, Technology & Society* 26 (6): 485–496.

Fisher, Erik, and Clark Miller. 2009. "Contextualizing the Engineering Laboratory." In *Engineering in Context*, ed. Steen H. Christensen, Bernard Delahousse, and Martin Meganck, 369–381. Aarhus: Academica.

Foucault, Michel. 1976. *La volonté de savoir*. Paris: Gallimard.

Foucault, Michel. 1984. *L'usage des plaisirs*. Paris: Gallimard.

Foucault, Michel. [1984] 2001a. "Le souci de la vérité." In *Dits et Ecrits II: 1976–1988*, 1487–1497. Paris: Gallimard.

Foucault, Michel. [1984] 2001b. "Polémique, politique et problématisations." In *Dits et Ecrits II: 1976–1988*, 1410–1417. Paris: Gallimard.

Foucault, Michel. [1971] 2001c. "Nietzche, la généalogie et l'histoire." In *Dits et Ecrits I: 1954–1975*, 1004–1024. Paris: Gallimard.

Foucault, Michel. [1984] 2001d. "Entretien avec C. Baker." In *Dits et Ecrits II: 1976–1988*, 1507–1515. Paris: Gallimard.

Foucault, Michel, and Gilles Deleuze. [1972] 2001. "Entretien: Les intellectuels et le pouvoir." In *Dits et Ecrits I: 1954–1975*, 1174–1183. Paris: Gallimard.

Fourniau, Jean-Michel. 2007. "L'expérience démocratique des 'citoyens en tant que riverains' dans les conflits d'aménagement." *Revue européenne des sciences sociales* 45 136 149–179.

Fraser, Nancy. 1990. "Rethinking the Public Sphere: A Contribution to the Critique of Actually Existing Democracy." *Social Text* 25/26:56–80.

Gaïti, Brigitte. 2007. "L'opinion publique dans l'histoire politique: Impasses et bifurcations." *Le Mouvement Social* 221:95–104.

Gavelin, Karin, Richard Wilson, and Rob Doubleday. 2007. *Democratic Technologies? The Final Report of the Nanotechnology Engagement Group (NEG)*. London: Involve.

Gomart, Emilie, and Maarten Hajer. 2003. "Is THAT Politics?" In *Social Studies of Science and Technology: Looking Back, Ahead*, ed. Bernward Joerges and Helga Nowotny, 33–61. Dordrecht: Kluwer.

Graber, Fabien. 2009. *'Paris a besoin d'eau': Projet, dispute et délibération technique dans la France napoléonienne*. Paris: CNRS Éditions.

Guston, David. 2004. "Forget Politicizing Science: Let's Democratize Science!" *Issues in Science and Technology* 21 (1): 25–28.

Guston, David. 2011. "Participating Despite Questions: Toward a More Confident Participatory Technology Assessment." *Science and Engineering Ethics* 17 (4): 691–697.

Guston, David, and Daniel Sarewitz. 2002. "Real-time Technology Assessment." *Technology in Society* 24:93–109.

Hamlett, Patrick. 2003. "Technology Theory and Deliberative Democracy." *Science, Technology & Human Values* 28 (1): 112–140.

Hamlett, Patrick, Michael Cobb, and David Guston. 2013. "National Citizens' Technology Forum: Nanotechnologies and Human Enhancement." In *Nanotechnology, the Brain, and the Future*, ed. Sean Hays, Jason Robert, Clark Miller, and Ira Bennett, 265–283. Dordrecht: Springer.

Hecht, Gabrielle. 1998. *The Radiance of France*. Cambridge, MA: MIT Press.

Hess, David. 2010. "Environmental Reform Organizations and Undone Science in the United States: Exploring the Environmental, Health and Safety Implications of Nanotechnology." *Science as Culture* 19 (2): 181–214.

Hilgartner, Stephen, and Charles Bosk. 1988. "The Rise and Fall of Social Problems: A Public Arena Approach." *American Journal of Sociology* 94 (1): 53–78.

Hubert, Matthieu. 2007. "Hybridations instrumentales et identitaires dans la recherche sur les nanotechnologies: Le cas d'un laboratoire public au travers de ses collaborations académiques et industrielles." *Revue d'Anthropologie des Connaissances* 2:243–266.

Hughes, James. 2004. *Citizen Cyborg: Why Democratic Societies Must Respond to the Redesigned Human of the Future*. New York: Basic Books.

Hullmann, Angela. 2006a. "Who Is Winning the Global Nanorace?" *Nature Nanotechnology* 1:81–83.

Hullmann, Angela. 2006b. *The Economic Development of Nanotechnology—An Indicator Based Analysis*. Brussels: European Commission.

Irwin, Alan. 2006. "The Politics of Talk." *Social Studies of Science* 36 (2): 299–320.

Jain, Kewal K. 2004. "Applications of Biochips: From Diagnostics to Personalized Medicine." *Current Opinion in Drug Discovery & Development* 7:285–289.

Jain, Kewal K. 2005. "Role of Nanobiotechnology in Developing Personalized Medicine for Cancer." *Technology in Cancer Research & Treatment* 4 (6): 645–650.

James, William. [1908] 1977. *A Pluralistic Universe*. Cambridge, MA: Harvard University Press.

Jasanoff, Sheila. 1990. *The Fifth Branch, Science Advisers as Policy-Makers*. Cambridge, MA: Harvard University Press.

Jasanoff, Sheila. 1992. "Science, Politics and the Renegotiation of Expertise at EPA." *Osiris* 7:192–217.

Jasanoff, Sheila. 1996. "Beyond Epistemology: Relativism and Engagement in the Politics of Science." *Social Studies of Science* 26 (2): 393–418.

Jasanoff, Sheila, ed. 2004. *States of Knowledge: The Coproduction of Science and Social Order.* London: Routledge.

Jasanoff, Sheila. 2005. *Designs on Nature: Science and Democracy in Europe and the United States.* Princeton: Princeton University Press.

Jasanoff, Sheila. 2011. "Constitutional Moments in Governing Science and Technology." *Science and Engineering Ethics* 17 (3): 621–638.

Jasanoff, Sheila. 2012. *Science and Public Reason.* London: Routledge.

Jobert, Arthur. 1998. "L'aménagement en politique: Ou ce que le syndrome NIMBY nous dit de l'intérêt général." *Politix* 11:67–92.

Joly, Pierre-Benoît, et al. 2005. *Démocratie locale et maîtrise sociale des nanotechnologies: Les publics grenoblois peuvent-ils participer aux choix scientifiques et techniques.* Rapport de la mission pour la Métro, September 22.

Joly, Pierre-Benoît, and Alain Kaufman. 2008. "Lost in Translation: The Need for Upstream Engagement with Nanotechnology on Trial." *Science as Culture* 17 (3): 225–247.

Joly, Pierre-Benoît, Brice Laurent, Claire Marris, and Douglas Robinson. 2011. *Biologie de Synthèse: Conditions d'un Dialogue avec la Société, rapport pour le Ministère de la Recherche et de l'Enseignement Supérieur.* Marne-La-Vallée: IFRIS.

Joly, Pierre-Benoît, and Claire Marris. 2001. "Agenda-Setting and Controversies: A Comparative Approach to the Case of GMOs in France and the United States." INRA-STEPE.

Jones, Richard. 2009. "Are You a Responsible Nanoscientist?" *Nature Nanotechnology* 4:336.

Joss, Simon, and Sergio Bellucci, eds. 2002. *Participatory Technology Assessment: European Perspectives.* London: Centre for the Study of Democracy.

Joss, Simon, and John Durant. 1995. *Public Participation in Science: The Role of Consensus Conference in Europe.* London: Science Museum.

Jouzel, Jean-Noël, and Pierre Lascoumes. 2011. "Le règlement REACH: Une politique européenne de l'incertain: Un détour de régulation pour la gestion des risques chimiques." *Politique européenne* 1:185–214.

Juengst, Eric. 1991. "The Human Genome Project and Bioethics." *Kennedy Institute of Ethics Journal* 1 (1): 71–74.

Juengst, Eric. 1994. "Human Genome Research and the Public Interest: Progress Notes from an American Science Policy Experiment." *American Journal of Human Genetics* 54:121–128.

Juengst, Eric. 1996. "Self-critical Federal Science? The Ethics Experiment within the U.S. Human Genome Project." *Social Philosophy & Policy* 13 (2): 63–95.

Kaiser, Mario, Monika Kurath, and Sabine Maasen, eds. 2010. *Governing Future Technologies—Nanotechnology and the Rise of an Assessment Regime*. Dordrecht: Springer.

Kearnes, Matthew, and Arie Rip. 2009. "The Emerging Governance Landscape of Nanotechnology." In *Jenseits von Regulierung: Zum politischen Umgang mit der Nanotechnologie*, ed. Stefan Gammel, Andreas Lösch, and Alfred Nordmann, 97–121. Berlin: Akademische Verlagsgesellschaft.

Kearnes, Matthew, and Brian Wynne. 2007. "On Nanotechnology and Ambivalence: The Politics of Enthusiasm." *NanoEthics* 1 (2): 131–142.

Kelty, Christopher. 2009. "Beyond Implications and Applications: The Story of Safety by Design." *NanoEthics* 3 (2): 79–96.

Khushf, George. 2004. "The Ethics of Nanotechnology: Vision and Values for a New Generation of Science and Engineering." In *Emerging Technologies and Ethical Issues in Engineering: Papers from a Workshop, October 14–15, 2003*, National Academy of Engineering, 29–56. Washington, DC: National Academies Press.

Khushf, George. 2007a. "Upstream Ethics in Nanomedicine: A Call for Research." *Nanomedicine* 2 (4): 511–521.

Khushf, George. 2007b. "The Ethics of NBIC Convergence." *Journal of Medicine and Philosophy* 32 (3): 185–196.

Khushf, George. 2007c. "Open Questions in the Ethics of Convergence." *Journal of Medicine and Philosophy* 32 (3): 299–310.

Khushf, George. 2009. "Open Evolution and Human Agency. The Pragmatics of Upstream Ethics in the Design of Artificial Life." In *The Ethics of Protocells: Moral and Social Implications of Creating Life in the Laboratory*, ed. Mark Bedau and Emily Park, 223–252. Cambridge, MA: MIT Press.

Kingdon, John. 1984. *Agendas, Alternatives, and Public Policies*. Boston: Little, Brown and Co.

Kleinman, Daniel, and Jason Delborne. 2009. "Engaging Citizens: The High Cost of Citizen Participation in High Technology." Paper presented at the Society for the Social Studies of Science annual meeting, Washington, DC, October 28–30.

Krimsky, Sheldon. 1984. "Beyond Technocracy: New Routes for Citizen Involvement in Social Risk Assessment." In *Citizen Participation in Science Policy*, ed. James Petersen, 43–61. Amherst: University of Massachussetts Press.

Kurath, Monika. 2009. "Nanotechnology Governance: Accountability and Democracy in New Modes of Regulation and Deliberation." *Science, Technology and Innovation Studies* 5 (2): 87–110.

Lacour, Stéphanie. 2012. "L'impossible définition des substances à l'état nanoparticulaire: Éléments d'analyse du décret n°2012-232 du 17 février 2012 relatif à la déclaration annuelle des substances à l'état nanoparticulaire pris en application de l'article L. 523-4 du code de l'environnement." *Revue Environnement et Développement Durable* 5:14–20.

Lascoumes, Pierre. 1994. *L'éco-pouvoir: Environnements et Politiques*. Paris: La Découverte.

Latour, Bruno. 1988a. "The Politics of Explanation: An Alternative." In *Knowledge and Reflexivity: New Frontiers in the Sociology of Knowledge*, ed. Steve Woolgar, 155–176. London: Sage.

Latour, Bruno. 1988b. *The Pasteurization of France, Followed by Irreductions*. Cambridge, MA: Harvard University Press.

Latour, Bruno. 1991. *We Have Never Been Modern*. Cambridge, MA: Harvard University Press.

Latour, Bruno. 1992. *Aramis, ou l'amour des techniques*. Paris: La Découverte.

Latour, Bruno. 2004. "Why Has Critique Run out of Steam? From Matters of Fact to Matters of Concern." *Critical Inquiry* 30:225–248.

Latour, Bruno. 2005. *Re-assembling the Social: An Introduction to Actor-Network Theory*. Oxford: Oxford University Press.

Latour, Bruno. 2007. "Turning around Politics: A Note on Gerard DeVries's Paper." *Social Studies of Science* 37 (5): 811–820.

Laurent, Brice. 2007. "Diverging Convergences. Competing Meanings of Nanotechnology and Converging Technologies in a Local Context." *Innovation: The European Journal of Social Science Research* 20 (4): 343–357.

Laurent, Brice. 2010. *Les politiques des nanotechnologies: Pour un traitement démocratique d'une science émergente*. Paris: Charles Léopold Mayer.

Laurent, Brice. 2011. "Technologies of Democracy: Experiments and Demonstrations." *Science and Engineering Ethics* 17 (3): 649–666.

Laurent, Brice. 2016a. "Perfecting European Democracy: Science as a Problem of Political and Technological Progress." In *Perfecting Human Futures: Technology,*

Secularization and Eschatology, ed. Benjamin Hurlbut and Hava Tirosh-Samuelson, 217–237. Dordrecht: Springer.

Laurent, Brice. 2016b. "Political Experiments that Matter: Ordering Democracy from Experimental Sites." *Social Studies of Science* 46 (5): 773–794.

Law, John. 2009. "Seeing Like a Survey." *Cultural Sociology* 3 (2): 239–256.

Law, John. 2010. "The Greer-Bush Test: On Politics in STS." In *Débordements: Mélanges offerts à Michel Callon*, ed. Madeleine Akrich, Yannick Barthe, Fabian Muniesa, and Philippe Mustar, 269–281. Paris: Presses des Mines.

Lefort, Claude. 1986. *Essais sur le politique*. Paris: Seuil.

Lemoine, Benjamin, and Yann LeLann. 2012. "Les comptes des générations: Les valeurs du futur et la transformation de l'Etat social." *Actes de la Recherche en Sciences Sociales* 194:78–101.

Levidow, Les. 1998. "Democratizing Technology—or Technologizing Democracy? Regulating Agricultural Biotechnology in Europe." *Technology in Society* 20 (2): 211–226.

Levidow, Les, and Susan Carr. 1997. "How Biotechnology Regulation Sets a Risk/ Ethics Boundary." *Agriculture and Human Values* 14:29–43.

Lewenstein, Bruce, and Rick Bonney. 2004. "Different Ways of Looking at Public Understanding of Research." In *Creating Connections: Museums and the Public Understanding of Current Research*, ed. David Chittenden, Graham Farmelo, and Bruce Lewenstein, 63–72. Walnut Creek, CA: Altamira Press.

Lezaun, Javier, and Linda Soneryd. 2007. "Consulting Citizens: Technologies of Elicitation and the Mobility of Publics." *Public Understanding of Science* 16:279–297.

Lin, Patrick. 2007. "In Defense of Nanoethics: A Reply to Adam Keiper." *The New Atlantis*, Correspondence (Summer).

Lösch, Andreas. 2006. "Anticipating the Futures of Nanotechnology: Visionary Images as Means of Communication." *Technology Analysis and Strategic Management* 18 (3–4): 393–409.

Lundvall, Bengt-Ake, and Edward Lorenz. 2011. "From the Lisbon Strategy to Europe 2020." In *Toward a Social Investment Welfare State? Ideas, Policies and Challenges*, ed. Nathalie Morel, Bruno Palier, and Jakob Palme, 333–351. Bristol: Policy Press.

Macdonald, Sharon, ed. 1996. *The Politics of Display*. London: Routledge.

Macnaghten, Phil, Matthew Kearnes, and Brian Wynne. 2005. "Nanotechnology, Governance and Public Deliberation: What Role for the Social Sciences?" *Science Communication* 27 (2): 268–287.

Mangematin, Vincent, Arie Rip, Aurélie Delemarle, and Douglas Robinson. 2005. "The Role of Regional Institutional Entrepreneurs in the Emergence of Clusters in Nanotechnology." Université Pierre Mendès-France/GAEL Working Paper.

Manin, Bernard. 1997. *Principles of Representative Government*. Cambridge: Cambridge University Press.

Marchant, Gary E., and Kenneth L. Mossman. 2004. *Arbitrary and Capricious: The Precautionary Principle in the European Union Courts*. Washington, DC: AEI Press.

Marchi, Florence, Daniela Urma, Sylvain Marlière, Jean-Loup Florens, Joël Chevrier, and Annie Luciani. 2005, "Educational Tool for Nanophysics Using Multisensory Rendering." Paper at the First Joint Eurohaptics Conference and Symposium on Haptic Interfaces for Virtual Environment and Teleoperator Systems, Pisa, Italy, March 18–20.

Marcus, George. 1995. "Ethnography in/of the World System: The Emergence of Multi-sited Ethnography." *Annual Review of Anthropology* 24:95–117.

Marlière, Sylvain, Daniela Urma, Jean-Loup Florens, and Florence Marchi. 2004. "Multi-sensorial Interaction with a Nano-scale Phenomenon: The Force Curve." Paper presented at the EuroHaptics conference, Munich, Germany, June 5–7.

Marres, Noortje. 2007. "The Issues Deserve More Credit: Pragmatist Contributions to the Study of Public Involvement in Controversy." *Social Studies of Science* 37 (5): 759–780.

Marres, Noortje. 2012. *Material Participation*. London: Palgrave.

Marris, Claire. 2001. *Final Report of the Public Attitudes to Biotechnology in Europe*. Lancaster: Centre for the Study of Environmental Change, Lancaster University.

Maynard, Andrew. 2011. "Don't Define Nanomaterials." *Nature* 475:31.

McCain, Lauren. 2002. "Informing Technology Policy Decisions: The US Human Genome Project's Ethical, Legal, and Social Implications Programs as a Critical Case." *Technology in Society* 24 (1–2): 111–132.

McCarthy, Elise, and Christopher Kelty. 2010. "Responsibility and Nanotechnology." *Social Studies of Science* 40 (3): 405–432.

McCombs, Maxwell, and Donald Shaw. 1972. "The Agenda-Setting Function of Mass Media." *Public Opinion Quarterly* 36 (2): 176–187.

McCray, Patrick. 2005. "Will Small Be Beautiful? Making Policies for Our Nanotech Future." *History and Technology* 21 (2): 177–203.

Merz, Martina, and Peter Biniok. 2010. "How Technological Platforms Reconfigure Science-Industry Relations: The Case of Micro- and Nanotechnology." *Minerva* 48 (2): 105–124.

Meyer, Morgan. 2009. "From 'Cold' Science to 'Hot' Research: The Texture of Controversy." CSI Working Paper N16. Centre de Sociologie de l'Innovation, Paris.

Michael, Mike. 2000. "Futures of the Present: From Performativity to Prehension." In *Contested Futures: A Sociology of Prospective Techno-Science*, ed. Nick Brown, Brian Rappert, and Andrew Webster, 21–42. Aldershot: Ashgate.

Miller, Peter, and Ted O'Leary. 2007. "Mediating Instruments and Making Markets: Capital Budgeting, Science and the Economy." *Accounting, Organizations and Society* 32:701–734.

Mody, Cyrus. 2011. *Instrumental Community: Probe Microscopy and the Path to Nanotechnology*. Cambridge, MA: MIT Press.

Mol, Annemarie. 1999. "Ontological Politics: A Word and Some Questions." In *Actor Network Theory and After*, ed. John Law and John Hassard, 74–89. Oxford: Blackwell.

Mol, Annemarie. 2002. *The Body Multiple: Ontology in Medical Practice*. Durham, NC: Duke University Press.

Moor, James, and John Weckert. 2004. "Nanoethics: Assessing the Nanoscale from an Ethical Point of View." In *Discovering the Nanoscale*, ed. David Baird, Alfred Nordmann, and Joachim Schummer, 301–310. Amsterdam: IOS Press.

Moor, James. 2005. "Why We Need Better Ethics for Emerging Technologies." *Ethics and Information Technology* 7 (3): 111–119.

Morris, Douglas. 2004. "Globalization and Media Democracy: The Case of Indymedia." In *Shaping the Network Society*, ed. Douglas Schuler and Peter Day, 325–352. Cambridge, MA: MIT Press.

Mouffe, Chantal. 2005. *On the Political*. London: Routledge.

National Research Council. 2006. *A Matter of Size: Triennial Review of the National Nanotechnology Initiative*. Washington, DC: NRC.

National Science and Technology Council. 2011. *National Nanotechnology Initiative Strategic Plan*. Washington, DC: NSTC.

Nordmann, Alfred. 2004. *Converging Technologies—Shaping the Future of European Societies*. Brussels: European Commission.

Nordmann, Alfred. 2009. "European Experiments." *Osiris* 24 (1): 278–302.

Organization for Economic Cooperation and Development (OECD). 2012. *Planning Guide for Public Engagement and Outreach in Nanotechnology: Key Points for Consideration When Planning Public Engagement Activities in Nanotechnology*. Paris: OECD.

Osborne, Thomas, and Nikolas Rose. 1999. "Do the Social Sciences Create Phenomena: The Case of Public Opinion Research." *British Journal of Sociology* 50 (3): 367–396.

Owen, Richard, Phil Macnaghten, and Jack Stilgoe. 2012. "Responsible Research and Innovation: From Science in Society to Science for Society, with Society." *Science & Public Policy* 39 (6): 751–760.

Pestre, Dominique. 1990. "Louis Néel, le magnétisme et Grenoble: Récit de la création d'un empire physicien dans la province française 1940–1965." *Cahiers pour l'histoire du CNRS* 8: 1–188.

Pestre, Dominique. 2004. "Thirty Years of Science Studies: Knowledge, Society and the Political." *History and Technology* 20 (4): 351–369.

Philbrick, Mark, and Javiera Barandiaran. 2009. "The National Citizens' Technology Forum: Lessons for the Future." *Science & Public Policy* 36 (5): 335–347.

Plomer, Aurora. 2008. "The European Group on Ethics: Law, Politics and the Limits of Moral Integration in Europe." *European Law Journal* 14 (6): 839–859.

PMO. 2008. *Aujourd'hui le Nanomonde. Nanotechnologies: Un projet de société totalitaire.* Paris: L'Échappée.

PMO. 2009. *À la recherché du nouvel ennemi.* Paris: L'Échappée.

Porter, Theodore. 1996. *Trust in Numbers: The Pursuit of Objectivity in Science and Public Life.* Princeton, NJ: Princeton University Press.

Powell, Maria, and Daniel Lee Kleinman. 2006. "Building Citizen Capacities for Participation in Nanotechnology Decision-making: The Democratic Virtues of the Consensus Conference Model." *Public Understanding of Science* 17 (3): 329–348.

Powell, Maria, Mathilde Colin, Daniel Lee Kleinman, Jason Delborne, and Ashley Anderson. 2011. "Imagining Ordinary Citizens? Conceptualized and Actual Participants for Deliberations on Emerging Technologies. *Science as Culture* 20 (1): 37–70.

President's Council on Bioethics. 2003. *Beyond Therapy: Biotechnology and the Pursuit of Happiness.* Washington, DC: The President's Council on Bioethics.

President's Council on Bioethics. 2008. *Human Dignity and Bioethics.* Washington, DC: The President's Council on Bioethics.

Prosseda, Kathleen. 2002. "Policy Debate on the Internet: Panelists Evaluate the Process." Paper presented at the International Symposium on Technology and Society, Social Implications of Information and Communication Technology Proceedings, Raleigh, NC, June 8.

Putnam, Hilary. 1989. "Objectivity and the Science/Ethics Distinction." Wider Working Paper No. 70, World Institute for Development Economics Research of the United Nations University, Helsinki, Finland.

Putnam, Hilary. 2004. *Ethics without Ontology.* Cambridge, MA: Harvard University Press.

Rabeharisoa, Vololona, and Michel Callon. 2004. "Patients and Scientists in French Muscular Dystrophy Research." In *States of Knowledge: The Co-Production of Science and Social Order,* ed. Sheila Jasanoff, 142–160. London: Routledge.

Rabinow, Paul. 2003. *Anthropos Today: Reflections on Modern Equipment.* Princeton, NJ: Princeton University Press.

Rajan, Kaushik S. 2006. *Biocapital: The Constitution of Postgenomic Life.* Durham, NC: Duke University Press.

Regis, Ed. 2004. "The Incredible Shrinking Man." *Wired* 12 (10). http://www.wired.com/2004/10/drexler/ (accessed July 20, 2016).

Reich, Christine, Larry Bell, Elizabeth Kollman, and Elissa Chin. 2007. "Fostering Civic Dialogue: A New Role for Science Museums?" *Museums and Social Issues* 2 (2): 207–220.

Reich, Christine, Julie Gross, Elizabeth Kunz Kollmann, Jane Morgan, and Amy Grack Nelson. 2011. *Review of NISE Network Evaluation Findings: Year 1-5.* NISE Network. http://www.informalscience.org/sites/default/files/Review_of_NISE_Net_Evaluation_Findings.pdf (accessed July 20, 2016).

Renn, Ortwinn, and Mihail Roco. 2006. "Nanotechnology and the Need for Risk Governance." *Journal of Nanoparticle Research* 8:153–191.

Revel, Martine, Cécile Blatrix, Loïc Blondiaux, Jean-Michel Fourniau, Bernard Heriard-Dubreil, and Rémi Lefebvre, eds. 2007. *Le débat public: Une expérience française de démocratie participative.* Paris: La Découverte.

Rip, Arie. 2006. "Folk Theories of Nanotechnologists." *Science as Culture* 15 (4): 349–365.

Rip, Arie. 2010. "Social Robustness and the Mode 2 Diagnosis." *Science, Technology and Innovation Studies* 6 (1): 71–74.

Robinson, Douglas, Arie Rip, and Vincent Mangematin. 2006. "Technological Agglomeration and the Emergence of Clusters and Networks in Nanotechnology." Université Pierre Mendès-France/GAEL Working Paper.

Rochefort, David, and Roger Cobb, eds. 1994. *The Politics of Problem Definition.* Lawrence: University Press of Kansas.

Roco, Mihail. 2004. "The US National Nanotechnology Initiative after 3 Years (2001–2003)." *Journal of Nanoparticle Research* 6:1–10.

Roco, Mihail. 2005. "Environmentally Responsible Development of Nanotechnology." *Environmental Science & Technology* 39 (5): 106–112.

Roco, Mihail, and William Bainbridge, eds. 2001. *Societal Implications of Nanoscience and Nanotechnology.* Dordrecht: Springer.

Roco, Mihail, and William Bainbridge, eds. 2003a. *Converging Technologies for Improving Human Performance: Nanotechnology, Biotechnology, Information Technology and Cognitive Science*. Dordrecht: Kluwer Academic Publishers.

Roco, Mihail, and William Bainbridge, eds. 2003b. *Societal Implications of Nanotechnology: Individual Perspectives*. Arlington, VA: National Science Foundation.

Roco, Mihail, and William Bainbridge. 2005. "Societal Implications of Nanoscience and Nanotechnology: Maximizing Human Benefit." *Journal of Nanoparticle Research* 7:1–13.

Roco, Mihail, Barbara Harthorn, David Guston, and Phil Shapira. 2011. "Innovative and Responsible Governance of Nanotechnology for Societal Development." In *Nanotechnology Research Directions for Societal Needs in 2020: Retrospective and Outlook*, ed. Mihail Roco, Chad Mirkin, and Mark Hersam, 561–617. Dordrecht: Springer.

Rosanvallon, Pierre. 1992. *Le sacre du citoyen: Histoire du suffrage universel en France*. Paris: Gallimard.

Rosanvallon, Pierre. 2008. *Counter-Democracy: Politics in the Age of Distrust*. Cambridge: Cambridge University Press.

Rosanvallon, Pierre. 2011a. *Democratic Legitimacy: Impartiality, Reflexivity, Proximity*. Princeton, NJ: Princeton University Press.

Rosanvallon, Pierre. 2011b. *La Société des Egaux*. Paris: Seuil.

Rose, Nikolas. 1999. *Powers of Freedom*. Cambridge: Cambridge University Press.

Rose, Nikolas. 2001. "The Politics of Life Itself." *Theory, Culture & Society* 18 (6): 1–30.

Rowe, Gene, and Lynn Frewer. 2000. "Public Participation Methods: A Framework for Evaluation." *Science, Technology & Human Values* 25 (1): 3–29.

Rowe, Gene, and Lynn Frewer. 2005. "A Typology of Public Engagement Mechanisms." *Science, Technology & Human Values* 30 (2): 251–290.

Ruivenkamp, Martin, and Arie Rip. 2011. "Entanglement of Imaging and Imagining of Nanotechnology." *NanoEthics* 5 (2): 185–193.

Sabel, Charles F., and Jonathan Zeitlin. 2003. "Active Welfare, Experimental Governance, Pragmatic Constitutionalism: The New Transformation of Europe." Paper presented at the International Conference of the Hellenic Presidency of the European Union, "The Modernisation of the European Social Model and EU Policies and Instruments," Ioannina, Greece, May 21–22.

Sachs, Noah. 2009. "Jumping the Pond: Transnational Law and the Future of Chemical Regulation." *Vanderbilt Law Review* 62:1817–1869.

Salter, Brian, and Mavis Jones. 2002. "Human Genetic Technologies, European Governance and the Politics of Bioethics." *Nature Reviews: Genetics* 3:808–814.

Schneider, Joseph W. 1985. "Social Problems Theory: The Constructionist View." *Annual Review of Sociology* 11:209–229.

Schot, Johan, and Arie Rip. 1997. "The Past and Future of Constructive Technology Assessment." *Technological Forecasting and Social Change* 54:251–268.

Schudson, Michael. 1998. *The Good Citizen: A History of American Civic Life*. New York: The Free Press.

Schummer, Joachim. 2005. "'Social and Ethical Implications of Nanotechnology': Meanings, Interest Groups, and Social Dynamics." *Technè* 8 (2): 56–87.

Selin, Cynthia. 2007. "Expectations and the Emergence of Nanotechnology." *Science, Technology & Human Values* 32 (2): 196–220.

Selin, Cynthia. 2009. "Diagnosing Futures: Producing Scenarios to Support Anticipatory Governance of Nanotechnology." Paper presented at the Society for the Social Studies of Science Annual Meeting, Washington, DC, October 28–31.

Shapin, Steven, and Simon Schaffer. 1985. *Leviathan and the Air-Pump: Hobbes, Boyle and the Experimental Life*. Princeton, NJ: Princeton University Press.

Snow, David. 1986. "Frame Alignment Processes, Micro-mobilization and Movement Participation." *American Sociological Review* 51 (4): 464–481.

Spector, Malcom, and John Kitsuse. 2001. *Constructing Social Problems*. Piscataway, NJ: Transaction Publishers.

Szyszczak, Erika. 2006. "Experimental Governance: The Open Method of Coordination." *European Law Journal* 12 (4): 486–502.

Talisse, Robert B., and Scott F. Aikin. 2005. "Why Pragmatists Cannot Be Pluralists." *Transactions of the Charles S. Peirce Society* 41 (1): 101–118.

Tallacchini, Mariachiara. 2006. "Politics of Ethics and EU Citizenship." *Politeia* 22 (83): 101–113.

Tallacchini, Mariachiara. 2009. "Governing by Values: Soft Tools, Hard Effects." *Minerva* 47 (3): 281–306.

Thireau, Isabelle, and Linshan Hua. 2013. *Les ruses de la démocratie: Protester en Chine*. Paris: Seuil.

Thoreau, François. 2013. "Embarquement immédiat pour les nanotechnologies responsables: Comment poser et re-poser la question de la réflexivité?" Unpublished PhD dissertation, Université de Liège.

Trubeck, David, and Trubeck Louise. 2005. "Hard and Soft Law in the Construction of Social Europe: The Role of the Open Method of Coordination." *European Law Journal* 11 (3): 343–364.

Tschudin, Verena. 2001. "European Experiences of Ethics Committees." *Nursing Ethics* 8 (2): 142–151.

Türk, Volker, Hugh Knowles, Holger Wallbaum, and Hans Kastenholz. 2006. *The Future of Nanotechnology: We Need to Talk*. Report of the Nanologue project, http://nanotech.law.asu.edu/Documents/2009/09/nanologue_scenarios_en_230_8727.pdf (accessed March 24, 2016).

U.S. Congress House of Representatives. 1992. *Designing Genetic Information Policy: The Need for an Independent Policy Review of the Ethical, Legal, and Social Implications of the Human Genome Project*. Committee on Government Operations, U.S. House of Representatives.

van Lente, Harro, and Arie Rip. 1998. "Expectations in Technological Developments: An Example of Prospective Structures to Be Filled in by Agency." In *Getting New Technologies Together*, ed. Cornelis Disco and Brend van der Meulen, 203–231. Berlin: de Gruyter.

van Oudheusden, Michiel, and Brice Laurent. 2013. "Shifting and Deepening Engagements: Experimental Normativity in Public Participation in Science and Technology." *Science, Technology & Innovation Studies* 9 (1): 1–21.

von Schomberg, René. 2009. "Organizing Collective Responsibility. On Precaution, Codes of Conduct and Understanding Public Debate." Keynote lecture at the meeting of the Society for the Study of Nanoscience and Emerging Technologies, Seattle, WA, September 11.

Wilsdon, James, and Rebecca Willis. 2004. *See-Through Science: Why Public Engagement Needs to Move Upstream*. London: Demos.

Winickoff, David, Sheila Jasanoff, Lawrence Busch, Robin Grove-White, and Brian Wynne. 2005. "Adjudicating the GM Food Wars: Science, Risk, and Democracy in World Trade Law." *Yale Journal of International Law* 30:81–123.

Winner, Langdon. 1980. "Do Artifacts Have Politics?" *Daedalus* 109 (1): 121–136.

Woodhouse, Edward, David Hess, Steve Breyman, and Brian Martin. 2002. "Science Studies and Activism: Possibilities and Problems for Reconstructivist Agendas." *Social Studies of Science* 32 (2): 297–319.

Woolf, Steven. 2008. "The Meaning of Translational Research and Why It Matters." *Journal of the American Medical Association* 299 (2): 211–213.

Wynne, Brian. 1996. "Misunderstood Misunderstandings: Social Identities and Public Uptake of Science." In *Misunderstanding Science? The Public Reconstruction of*

Science and Technology, ed. Alan Irwin and Brian Wynne, 19–46. Cambridge: Cambridge University Press.

Wynne, Brian. 2001. "Creating Public Alienation: Expert Cultures of Risk and Ethics on GMOs." *Science as Culture* 10:445–481.

Xperiment! (Bernd Kraeftner, Judith Kroell, and Isabel Warner). 2007. "Walking on a Story Board, Performing Shared Incompetence. Exhibiting 'Science' in the Public Realm." In *Exhibition Experiments: New Interventions in Art History*, ed. Sharon Macdonald and Paul Basu, 109–131. Oxford: Blackwell.

Yaneva, Albena, Tania Mara Rabesandratana, and Birgit Greiner. 2009. "Staging Scientific Controversies: A Gallery Test on Science Museums' Interactivity. *Public Understanding of Science* 18 (1): 79–90.

Yankelovitch, Daniel. 2001. *The Magic of Dialogue: Transforming Conflict into Cooperation*. New York: Simon and Schuster.

Youtie, Jan, Alan Porter, Philip Shapira, Li Tang, and Troy Benne. 2011. "The Use of Environmental, Health and Safety Research in Nanotechnology Research." *Journal of Nanoscience and Nanotechnology* 11 (1): 158–166.

Index

Inside Technology

edited by Wiebe E. Bijker, W. Bernard Carlson, and Trevor Pinch

Cyrus Mody, *The Long Arm of Moore's Law: Microelectronics and American Science*

Harry Collins, Robert Evans, and Christopher Higgins, *Justice, Accuracy, and Uncertainty: Technology's Attack on Umpires and Referees and How to Fix It*

Tiago Saraiva, *Fascist Pigs: Technoscientific Organisms and the History of Fascism*

Teun Zuiderent-Jerak, *Situated Intervention: Sociological Experiments in Health Care*

Basile Zimmermann, *Technology and Cultural Difference: Electronic Music Devices, Social Networking Sites, and Computer Encodings in Contemporary China*

Andrew J. Nelson, *The Sound of Innovation: Stanford and the Computer Music Revolution*

Sonja D. Schmid, *Producing Power: The Pre-Chernobyl History of the Soviet Nuclear Industry*

Casey O'Donnell, *Developer's Dilemma: The Secret World of Videogame Creators*

Christina Dunbar-Hester, *Low Power to the People: Pirates, Protest, and Politics in FM Radio Activism*

Eden Medina, Ivan da Costa Marques, and Christina Holmes, editors, *Beyond Imported Magic: Essays on Science, Technology, and Society in Latin America*

Anique Hommels, Jessica Mesman, and Wiebe E. Bijker, editors, *Vulnerability in Technological Cultures: New Directions in Research and Governance*

Amit Prasad, *Imperial Technoscience: Transnational Histories of MRI in the United States, Britain, and India*

Charis Thompson, *Good Science: The Ethical Choreography of Stem Cell Research*

Tarleton Gillespie, Pablo J. Boczkowski, and Kirsten A. Foot, editors, *Media Technologies: Essays on Communication, Materiality, and Society*

Catelijne Coopmans, Janet Vertesi, Michael Lynch, and Steve Woolgar, editors, *Representation in Scientific Practice Revisited*

Rebecca Slayton, *Arguments that Count: Physics, Computing, and Missile Defense, 1949–2012*

Stathis Arapostathis and Graeme Gooday, *Patently Contestable: Electrical Technologies and Inventor Identities on Trial in Britain*

Jens Lachmund, *Greening Berlin: The Co-Production of Science, Politics, and Urban Nature*

Chikako Takeshita, *The Global Biopolitics of the IUD: How Science Constructs Contraceptive Users and Women's Bodies*

Cyrus C. M. Mody, *Instrumental Community: Probe Microscopy and the Path to Nanotechnology*

Morana Alač, *Handling Digital Brains: A Laboratory Study of Multimodal Semiotic Interaction in the Age of Computers*

Gabrielle Hecht, editor, *Entangled Geographies: Empire and Technopolitics in the Global Cold War*

Michael E. Gorman, editor, *Trading Zones and Interactional Expertise: Creating New Kinds of Collaboration*

Matthias Gross, *Ignorance and Surprise: Science, Society, and Ecological Design*

Andrew Feenberg, *Between Reason and Experience: Essays in Technology and Modernity*

Nelly Oudshoorn and Trevor Pinch, editors, *How Users Matter: The Co-Construction of Users and Technology*

Peter Keating and Alberto Cambrosio, *Biomedical Platforms: Realigning the Normal and the Pathological in Late-Twentieth-Century Medicine*

Paul Rosen, *Framing Production: Technology, Culture, and Change in the British Bicycle Industry*

Maggie Mort, *Building the Trident Network: A Study of the Enrollment of People, Knowledge, and Machines*

Donald MacKenzie, *Mechanizing Proof: Computing, Risk, and Trust*

Geoffrey C. Bowker and Susan Leigh Star, *Sorting Things Out: Classification and Its Consequences*

Charles Bazerman, *The Languages of Edison's Light*

Janet Abbate, *Inventing the Internet*

Herbert Gottweis, *Governing Molecules: The Discursive Politics of Genetic Engineering in Europe and the United States*

Kathryn Henderson, *On Line and On Paper: Visual Representation, Visual Culture, and Computer Graphics in Design Engineering*

Susanne K. Schmidt and Raymund Werle, *Coordinating Technology: Studies in the International Standardization of Telecommunications*

Marc Berg, *Rationalizing Medical Work: Decision-Support Techniques and Medical Practices*

Eda Kranakis, *Constructing a Bridge: An Exploration of Engineering Culture, Design, and Research in Nineteenth-Century France and America*

Paul N. Edwards, *The Closed World: Computers and the Politics of Discourse in Cold War America*

Donald MacKenzie, *Knowing Machines: Essays on Technical Change*

Wiebe E. Bijker, *Of Bicycles, Bakelites, and Bulbs: Toward a Theory of Sociotechnical Change*

Louis L. Bucciarelli, *Designing Engineers*

Geoffrey C. Bowker, *Science on the Run: Information Management and Industrial Geophysics at Schlumberger, 1920–1940*

Wiebe E. Bijker and John Law, editors, *Shaping Technology / Building Society: Studies in Sociotechnical Change*

Stuart Blume, *Insight and Industry: On the Dynamics of Technological Change in Medicine*

Donald MacKenzie, *Inventing Accuracy: A Historical Sociology of Nuclear Missile Guidance*

Pamela E. Mack, *Viewing the Earth: The Social Construction of the Landsat Satellite System*

H. M. Collins, *Artificial Experts: Social Knowledge and Intelligent Machines*

http://mitpress.mit.edu/books/series/inside-technology